Recurrent Erosion Syndrome and Epithelial Edema

Helena M. Tabery

Recurrent Erosion Syndrome and Epithelial Edema

In Vivo Morphology in the Human Cornea

Springer

Helena M. Tabery, MD
Malmö
Sweden
helena.tabery@telia.com

Formerly
Ögonkliniken UMAS
205 02 Malmö
Sweden

ISBN 978-3-319-06544-1 ISBN 978-3-319-06545-8 (eBook)
DOI 10.1007/978-3-319-06545-8
Springer Cham Heidelberg New York Dordrecht London

Library of Congress Control Number: 2014956098

Printed on acid-free paper

Springer is part of Springer Science+Business Media (www.springer.com)

Preface

Part I of this book is dedicated to the *recurrent erosion syndrome*, a very common disturbance of the corneal surface causing symptoms that range between ocular discomfort and severe, disabling pain. The symptoms, whether starting spontaneously or after a minor injury, recur at irregular intervals. Their unpredictability often causes considerable stress to the individual affected.

Recurrent erosions occur both in apparently normal corneae and in corneae showing dystrophic changes of which the epithelial basement membrane dystrophy (EBMD, also termed map-dot-fingerprint or Cogan's microcystic dystrophy) is the most common one. In this book, recurrent erosions are treated as a syndrome with no ambition to subdivide them, link them to a particular underlying cause, or solve the question of a possible common denominator.

The images presented in this book cover a *broad spectrum of epithelial changes* of which some are manifestations of ongoing underlying *pathological process* while others indicate epithelial activity aiming at elimination of abnormal elements and *repair* of damage. Although some of the phenomena per se hint at their underlying mechanisms, the *dynamics* of the interplay between these two opposing forces captured in sequential photographs – particularly in those taken at 24 h intervals – are invaluable in the interpretation. They are included in the *cases* section that partly illustrates the typical features of the disease with phenomena present either in isolation or mixed in various combinations, and partly documents the variability of symptoms and findings in the same individual during various periods of time. Included are also images of the appearance, and the dynamics, of corneal *stromal infiltrates,* which is a rare but potentially sight-threatening *complication.*

Part II treats partly typical features of corneal *epithelial edema* and partly an occasional contemporaneous occurrence, and dynamics, of some phenomena indistinguishable from those commonly seen in the recurrent erosion syndrome.

This book differs from current textbooks in that the in vivo images captured at a higher *magnification level* than that used in standard photography additionally reveal details of epithelial changes, and in that the *non-touch approach* allows sequential observations of these details without disturbing the already deranged surface. The captured dynamics visualise the origin of several phenomena frequently encountered in the recurrent erosion syndrome and elucidates mechanisms behind them; some less common phenomena, however, remain unexplained.

Considering the terminology, I have adhered to that originating in older works but still currently used and added morphological descriptions of some details.

Finally, many *treatment* modalities have been tried but an evaluation of the results is difficult because of large individual variations. This book contains no treatment recommendations but the used treatment is mentioned wherever pertinent. There are several indications of hereditary factors in the literature but the mechanisms behind the pathology remain unknown. The morphology presented in this book reflects its sequelae and the response of natural reparative forces, makes their clinical appearance understandable, and facilitates differential diagnostics because the phenomena per se are well discernible within the range of the magnification power of a standard slit lamp.

Malmö, Sweden Helena M. Tabery, MD

Contents

About This Book

The photographs presented in this book have been chosen to show

Part I: Corneal Recurrent Erosion Syndrome

- The in vivo morphology of epithelial changes in recurrent erosions (Chap. 1)
- The in vivo dynamics of epithelial changes in recurrent erosions (Chap. 2)
- Clinical Applications (Chap. 2)
- The in vivo morphology and dynamics of stromal infiltrates in recurrent erosions (Chap. 3)

Part II: Corneal Epithelial Edema

- The in vivo morphology of corneal epithelial edema (Chap. 4)
- Rare additional phenomena and their dynamics (Chaps. 4 and 5)
- Clinical applications (Chap. 5)

The photographs were taken by *non-contact in vivo photomicrography*, a method that requires neither contact with the epithelium nor the use of anaesthetics. By this method are visualized structures that optically differ from their regularly organized surroundings; a normal corneal epithelium or stromal cells cannot be discerned. As there is no contact with the ocular surface, the architecture of epithelial changes is not disturbed by the examination, and there is no risk of spreading infections. The technique allows the use of various illumination modes to complement each other and a free application of *diagnostic dyes* to expand the information, e.g., 1 % fluorescein sodium and 1 % rose bengal (preservative-free solutions). These dyes are commonly used in clinical practice.

The *bars indicate 200 µm throughout the book.*

Abbreviations

EBMD	Epithelial basement membrane dystrophy
Fluorescein	Fluorescein sodium
KCS	Keratoconjunctivitis sicca
NaCl	Sodium chloride

Part I

Corneal Recurrent Erosion Syndrome

About the Corneal Recurrent Erosion Syndrome

Recurrent erosion is a common affection of the corneal epithelial surface causing *symptoms* that range between ocular discomfort, grittiness or foreign body sensation, and a sharp, disabling pain. Typically, the symptoms appear on awakening, and many patients learn to open their eyes carefully. A very common history are repeated early morning symptoms that subside during a couple of hours; after a few days, they either disappear entirely or become severe. The symptomatic episodes recur at irregular intervals the duration of which varies largely, and they are usually unilateral; in some individuals, the first involved eye remains quiet for a long time while symptoms appear in the fellow eye. During the quiet periods, the epithelium may appear normal or show changes relatable to the condition; a centrally located epithelial disturbance may cause irregular astigmatism resulting in fluctuating vision.

In many patients, symptoms are triggered by a per se trivial mechanical trauma most commonly caused by fingernails, sharp plant leaves, tree branches, sheets of paper, and similar objects. Usually, the original symptoms disappear rapidly only to reappear after a short period of time (within a couple of weeks). In some patients, however, recurrent erosions start without any apparent reason. Some patients report close relatives with similar problems.

Clinically, recurrent erosions occur in apparently normal corneae, corneae with cystic changes, corneae additionally showing features compatible with alterations of the basement membrane and/or deposits of plaque-like subsurface material, and corneae showing extensive changes relatable to epithelial basement membrane dystrophy (EBMD, also termed Cogan's microcystic dystrophy or map-dot-fingerprint dystrophy). In this dystrophy, not very common but probably the most common one of the corneal dystrophies, recurrent erosions may or may not occur. When located in the visual axis, the dystrophy may cause irregular astigmatism and visual impairment. Its hereditary features have been reported in the literature. Recurrent erosions occur also in other, rather rare, corneal dystrophies such as for example Reis-Bückler's and Meesman's dystrophy.

Recurrent erosions may resolve spontaneously or represent a serious problem. Several treatment modalities have been attempted (such as lubricating or/and hypertonic ointments and drops, topical steroids and tetracycline p.o, contact lenses, and surgical interventions) but at present there is no treatment obviously superior to the others.

Finally, it has to be noted that the well-established term 'recurrent erosion' encompasses both frank erosions (in the classic meaning of loss of substance) and other epithelial disturbances typical of the syndrome.

1 The Morphology of Recurrent Erosions

The morphological in vivo features of the recurrent erosion syndrome are relatable to various events: (1) manifestations of *ongoing pathological process* within the epithelium (partly as changes relatable to the epithelial cells themselves, and partly as changes implying alterations of the epithelial basement membrane and/or plaque-like deposits of abnormal sub- or intraepithelial material), (2) *disruptions of surface integrity* (epithelial disruptions within edematous areas and frank erosions) and (3) manifestations of *ongoing reparative process*. All may occur anywhere on the cornea, but the most common sites are its central and lower parts.

1. *Ongoing pathological process*. Of the changes relatable to *epithelial cells*, the smallest discernable entities are *rounded bodies*[1] of about 9–13 μm in diameter that for the most part are strongly light-reflecting. Such bodies are present individually, grouped or concentrated in larger areas and also within many cysts. With fluorescein, the light-reflecting ones appear brighter than their surroundings. Some larger rounded bodies may represent incipient cystic changes. Larger entities are *small, rounded cysts*,[2] of about 18–60 μm in diameter, some light-reflecting and others clear and many containing one or several rounded bodies. Variously large cysts may be present in the same area, individual or grouped; some close to each other may show propensity to confluence. Some stain green with fluorescein, others do not. Even larger entities are well-delineated *opaque cysts*,[3] more or less strongly light-reflecting (greyish to white) and dense in retroillumination, some rounded or oval (measuring about 50–130 μm in diameter), some long and wavy (the longest captured ones measuring >1 mm in length), and some with bizarre forms. Their contents are granular, and, with very rare exceptions (Fig. 5.15), they do not stain with fluorescein. Their surroundings appear elevated. Changes indicating *alterations of the epithelial basement membrane and/or plaque-like sub- or intraepithelial deposits* of abnormal material appear (a) as variously shaped *fine lines* best visible in retroillumination, some arranged in patterns reminding of fingerprints (also termed fingerprint lines) and others irregular

[1] By the present method, in retroillumination, cells with altered membranes and/or contents, incipient cell swelling, invading inflammatory cells and cell debris appear similar. In focal illumination, all with the exception of swelling cells with clear contents are light-reflecting, but as this difference is difficult to consistently capture, I have chosen to adhere to the term rounded bodies throughout. Their suggested interpretation in some individual cases indicates only a probable alternative. See also Final Remark.

[2] Because the term 'cysts', originating in clinical observations and descriptions, is well established I have adhered to it; the literature contains numerous histological examinations revealing cystic spaces (pseudocysts) containing either clear fluid or degenerating cells/cell debris, or both.

[3] See footnote 2.

H.M. Tabery, *Recurrent Erosion Syndrome and Epithelial Edema: In Vivo Morphology in the Human Cornea*, DOI 10.1007/978-3-319-06545-8_1, © Springer International Publishing Switzerland 2015 or straight; (b) as variously shaped strongly light-reflecting *thicker lines*, occasionally isolated but more often present at the edges of *greyish membranes* with irregular geographic shapes (also termed 'maps') which may contain darker 'holes'; and (c) *bleb-like* formations measuring about 15–140 μm in diameter and often confluent, transparent, clearly discernible only in retroillumination and only rarely causing slight surface elevations.

2. *Disruptions of surface integrity* occur partly as *breaks in the epithelial layer* within large (several mm in diameter) well-delineated, rounded or oval elevated areas of edematous epithelium, and partly as *frank erosions* showing epithelial detachment and loss of substance. In both, fluorescein diffuses into the tissues.
3. *Ongoing reparative process* is discernible as variously curved edges or *tips of epithelial sheets*. Depending on the stage of development, such tips may be present individually, close to each other, or form smooth lines at the edges of well-defined lesions. Opposite edges in apposition before merging show as superficial *epithelial lines* ('contact lines'), some smooth and others interrupted by fusiform expansions often containing rounded bodies; in some lines form pear-shaped cysts. The epithelium peripherally from the edges may appear smooth or contain rounded bodies and cysts. Fluorescein is pooling between the edges, in front of them and within the expansions and cysts. The surface peripherally from the edges often appears elevated.

At presentation, the cornea may show one or several of the basic phenomena in various combinations, and that per se or in combination with areas of epithelial disturbance, here termed *lesions*, measuring about 90 μm to 1.3 mm in their largest diameter. The lesions are either partly or wholly light-reflecting and show innumerable shapes. For the most part, they are sharply delineated by smooth, wavy edges, but occasionally the edges are irregular and show excrescences-like protrusions. The *smaller*, simple lesions usually contain many light-reflecting rounded bodies and/or granular material suggestive of cell debris; occasionally, some appear as light-reflecting areas of heaped-up rounded bodies reminiscent of epithelial infiltrates (such as in adenovirus or Thygesson's epithelial keratitis). The contents of the *larger* ones, complex and chaotic, comprise a mixture of elements relatable not only to disease process, such as various numbers of rounded bodies and/or cell debris, but also to reparative activity such as curved edges of epithelial sheets advancing into the disturbed areas. Notable are also smooth *rounded structures/islands* (of about 30–280 μm in diameter) that, with fluorescein, appear protruding; they are surrounded by lines in which the dye is pooling (cf. Figs. 1.29, 1.30, 1.31 and 1.32). A different phenomenon are circular (of about 80–100 μm in diameter, rarely up to 250 μm) or slightly oval (of about 100–160 μm in their largest diameter) areas that appear as if punched-out; they show rounded bodies (cf. Figs 1.33 and 1.34) or other irregularities and fluorescein is pooling in them (cf. Figs. 1.43 and 2.25).

With *fluorescein*, some lesions appear elevated (dark in the green stained tear film), while others stain partly or wholly green. In some, the visibility of the green staining may be hampered by strong light reflections from the lesions' contents. Complex lesions show various shades of green. The staining is always confined to the lesions and there is no diffusion into the surroundings. For dynamics of fluorescein staining of lesions, see Chap. 2, Cases 4 and 6, and for some additional interpretation of staining relevant to recurrent erosions, see Final Remark.

Rose bengal was used only in few patients in connection with KCS tests. None of them showed any corneal epithelial changes relatable to KCS, and the dye revealed only some damaged surface elements in the areas of epithelial disturbance occurring in the recurrent erosion syndrome. In the absence of diagnostic value, the task was not further pursued.

The *dynamics* and interplay of the destructive and reparative forces are shown in Case Reports (Chap. 2).

Fluorescein Staining (1)

Surface Elevations and Diffusion of the Dye

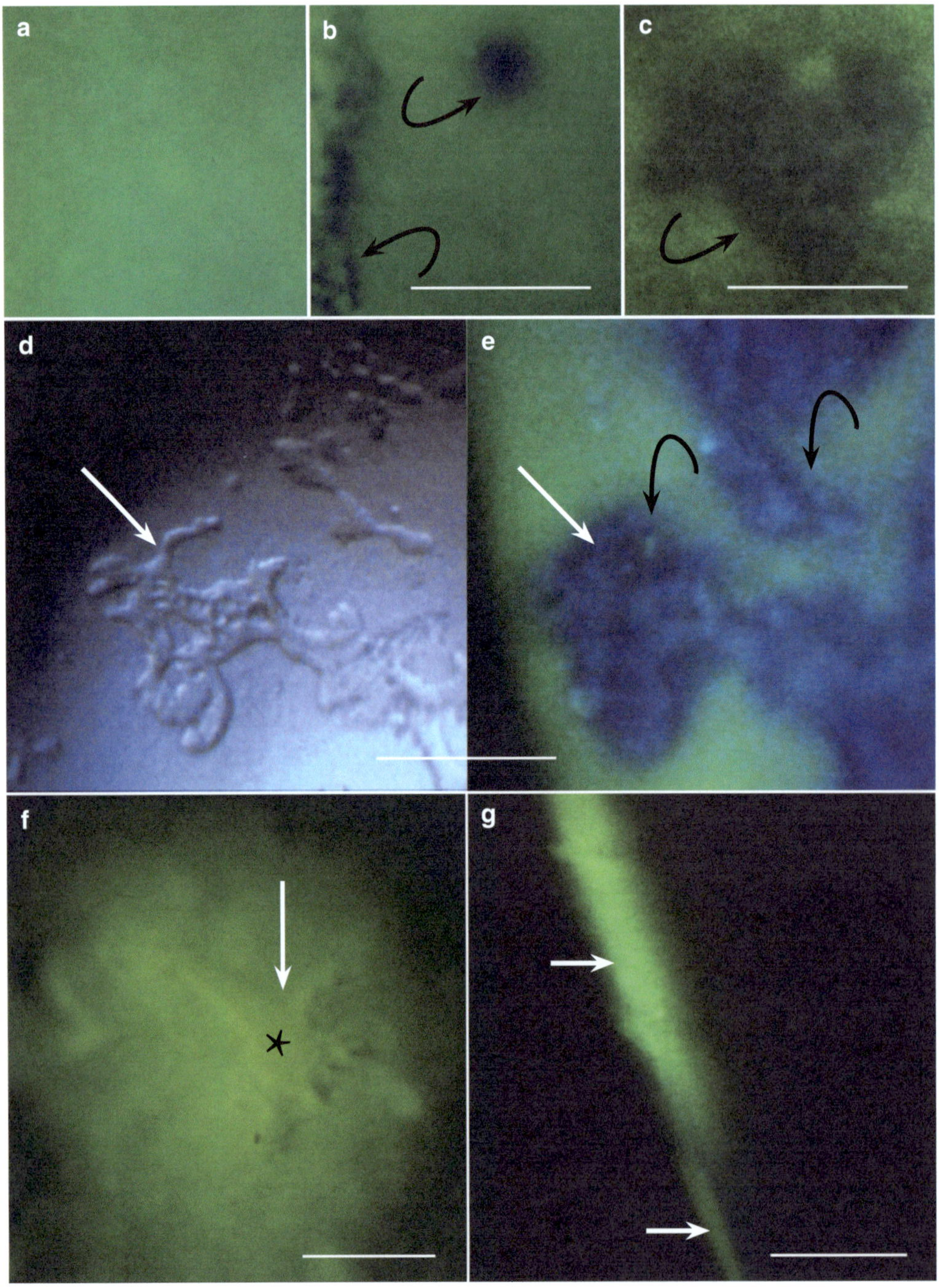

Fig. 1.1 (a) Normal precorneal tear film stained evenly green with fluorescein. (b–c) Surface elevations (*arrows*) appear dark. (d–e) show abnormal epithelial structures (*white arrows*) located below an elevated (dark, *black arrows*) but not disrupted surface (cf. Chap. 3, Case 1). (f) Fluorescein diffusion (green) around a surface disruption (*arrow*); its centre is indicated by (*asterisk*). (g) Fluorescein diffusion (*upper arrow*) into the surroundings captured in a narrow light beam (*lower arrow*)

Fluorescein Staining (2)

Pooling of the Dye

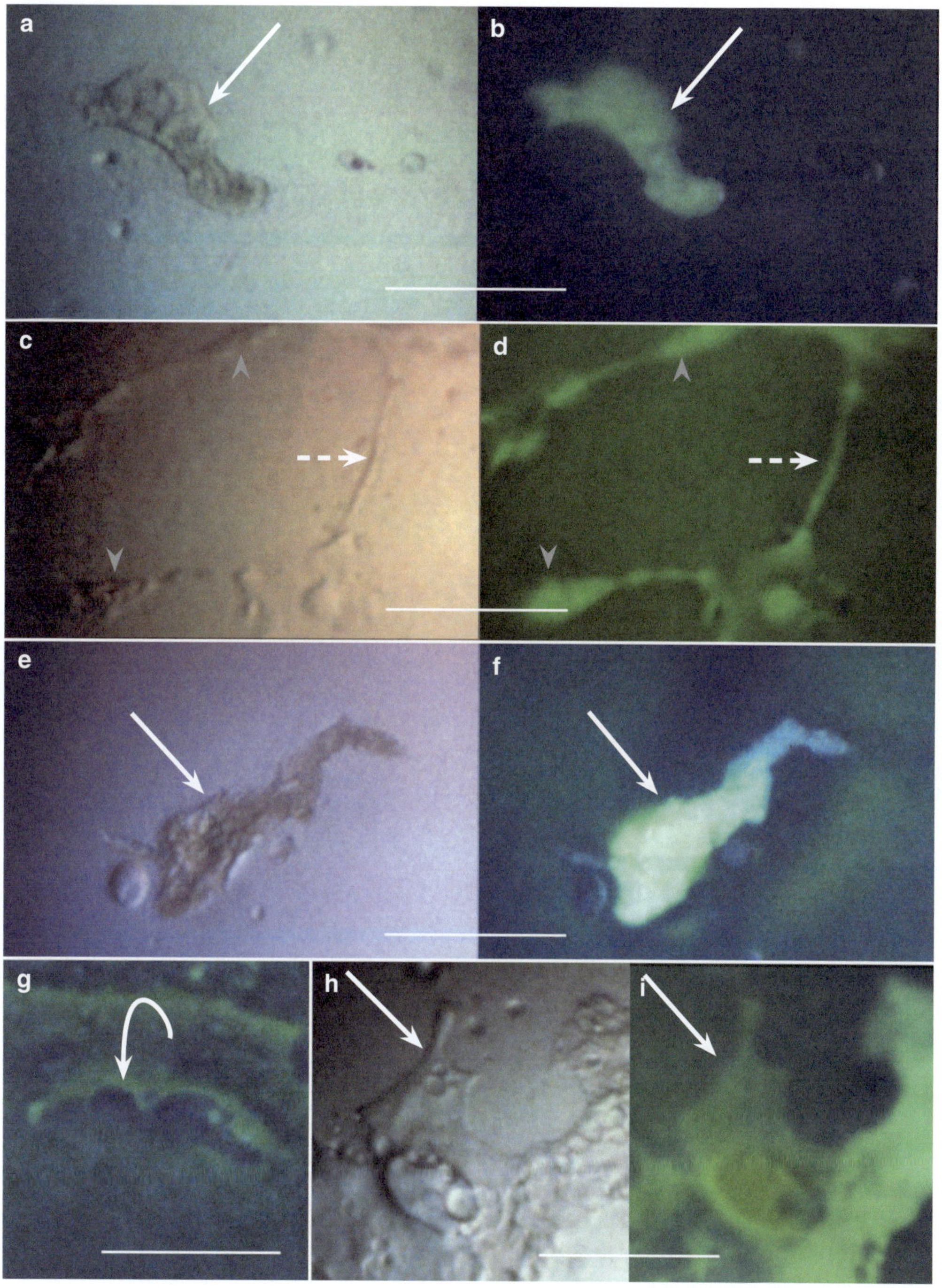

Fig. 1.2 *Fluorescein pooling in*: (**a–b**) a sharply delineated space (*arrows*); (**c–d**) epithelial lines (*arrows*) and variously shaped spaces (*arrowheads*); (**e–f**) a sharply delineated space (*arrows*) in which fluorescein visibility is partly hampered by a strong light reflection; (**g**) a relative surface depression (*arrow*); (**h–i**) a complex lesion (*arrows*). (The markers are placed in corresponding locations in each pair of photographs. **a**, **c**, **e** and **h**: before staining.) For dynamics of the green staining see Figs. 2.9, 2.11 and 2.14

Signs of Pathology

Rounded Bodies and Small Rounded Cysts (1)

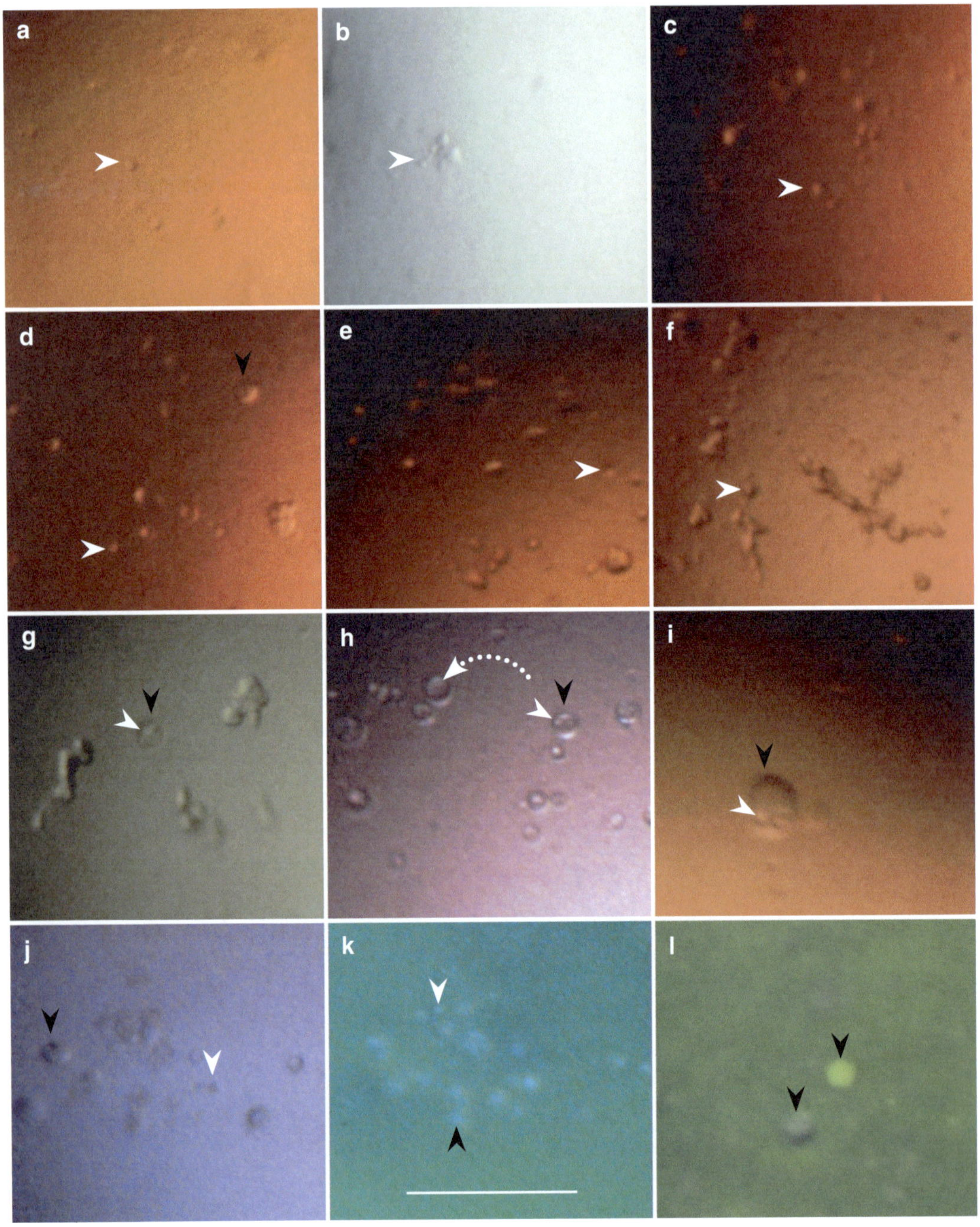

Fig. 1.3 (**a–k**) Rounded bodies (*white arrowheads*), individual or grouped, and small rounded cysts (*black arrowheads*) of which many contain rounded bodies (**g–i**, *white arrowheads*) and some a smooth rounded structure (**h**, *arrow*). (**j–k**) Many of the rounded bodies (*white arrowheads*) and cysts (*black arrowheads*) are light-reflecting (not the same area). (**l**) Of the two rounded cysts (*black arrowheads*) only the upper one stains green with fluorescein (for additional examples of fluorescein staining of rounded cysts, see Figs. 1.14, 1.32, 2.6 and 2.45)

Rounded Bodies and Small Rounded Cysts (2)

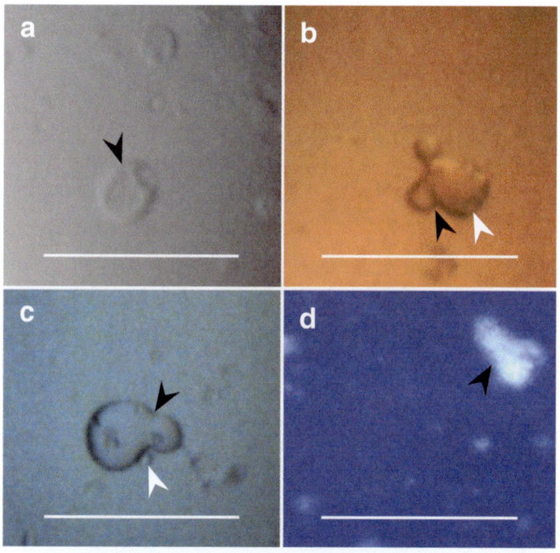

Fig. 1.4 (a–c) Epithelial cysts (*black arrowheads*), in apposition or merging (for fluorescein staining of similar cysts cf. Fig. 1.14). Rounded bodies (*white arrowheads*) are visible within the cysts and in the surroundings. (d) Also this light-reflecting cyst implies the result of confluence

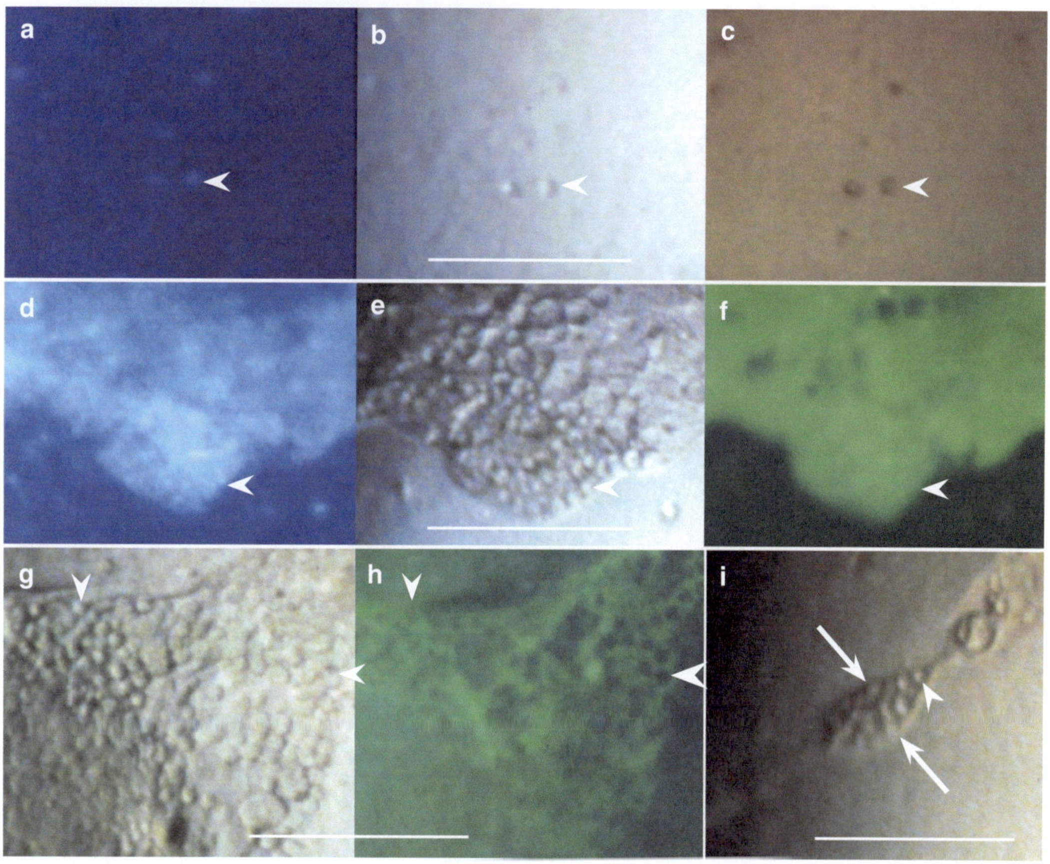

Fig. 1.5 (a–c) Rounded bodies (*arrowheads*) captured in different illumination modes. (d) A lesion the whitish appearance of which is partly due to (e) many light-reflecting rounded bodies (*arrowheads*) that (f) are difficult to discern after fluorescein staining. (g) Many rounded bodies (*arrowheads*) in (h) appearing as dark dots in the green stained fluid. (i) Rounded bodies (*arrowhead*) present in a fusiform expansion between approaching edges (*arrows*) of epithelial sheets (the arrowheads are placed in corresponding locations in each set of photographs)

Signs of Pathology

Opaque Cysts (1)

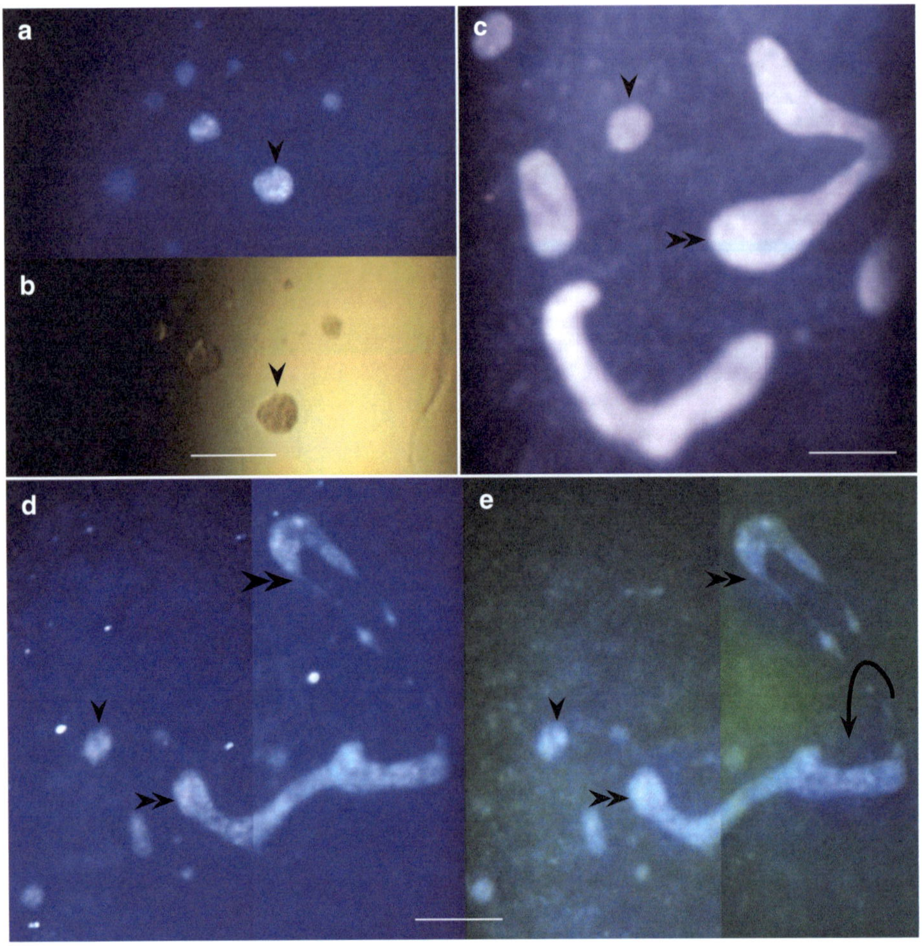

Fig. 1.6 (a) Rounded, variously large and strongly light-reflecting opaque cysts (*arrowheads*) that appear (b) dense (*dark*) in retroillumination. (c–e) Similar cysts, rounded (*arrowheads*) or variously shaped (*double arrowheads*) (cf. Fig. 1.7, opposite page). (d) The cysts do not (e) stain with fluorescein; the surrounding epithelium shows surface elevations (dark in the green stained tear film, *arrow*). See also Chap. 5, Case 3. (The arrowheads in (a, b) and (d, e) are placed in corresponding locations)

Comment

The photographs shown in Fig. 1.6 were captured in three patients with EBMD:

(a, b) In a 52-year-old myopic (−7.0) woman with fluctuating vision and glare for 7 months; both corneae showed centrally located light-reflecting opaque cysts and irregular astigmatism.

(c) As accidental finding in the upper cornea in a 60-year-old woman presenting with iridocyclitis.

(d, e) In a 46-year-old woman with a 6-month history of fluctuating vision in the right eye; the right cornea showed centrally located opaque cysts and irregular astigmatism and the left one similar cysts out of the visual axis.

In all three patients, both corneae additionally showed lines and maps. None of them had recurrent erosion symptoms.

Opaque Cysts (2)

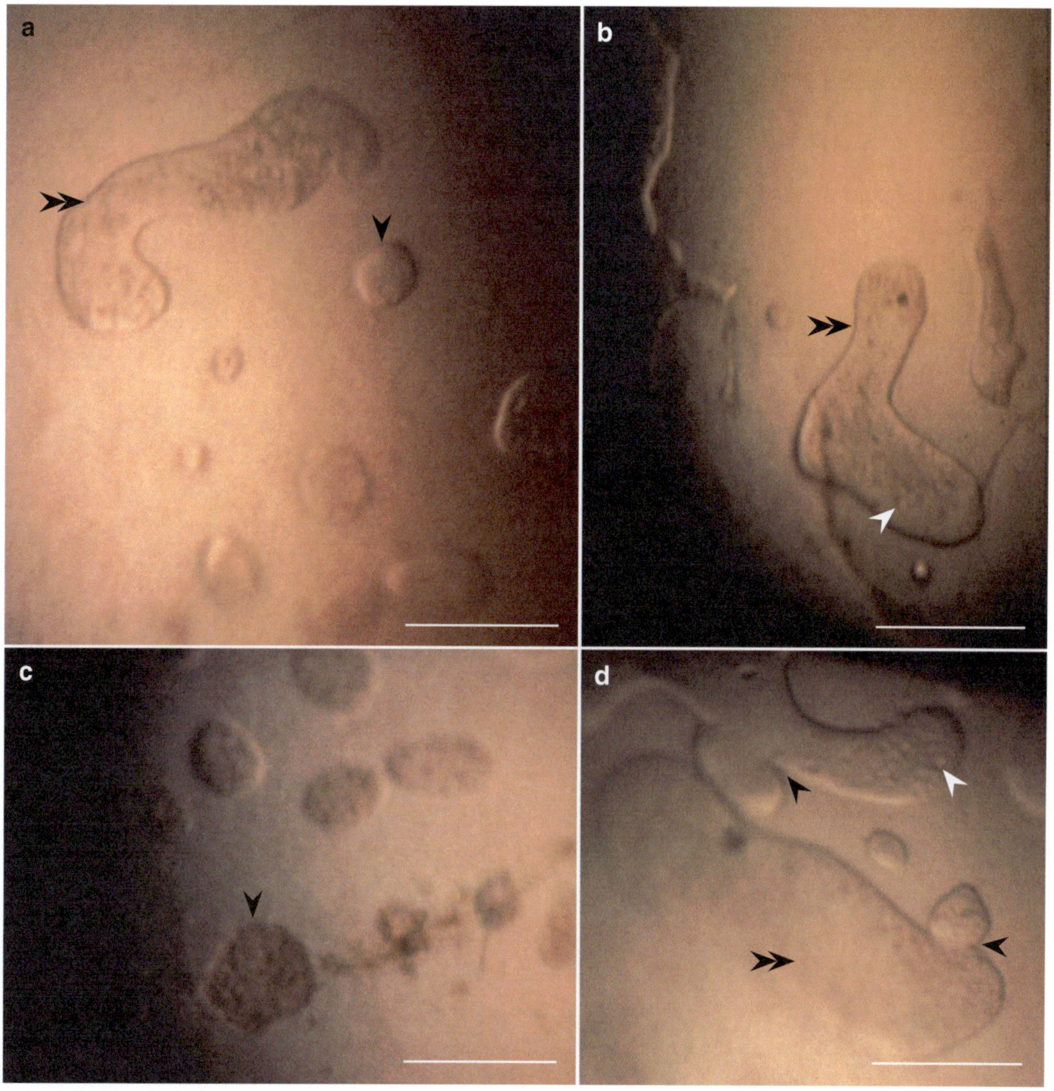

Fig. 1.7 (**a–d**) Rounded (*black arrowheads*) and variously shaped (*double arrowheads*) cysts in many of which is discernible granular material (rounded bodies and/or cell debris, *white arrowheads*). In (**d**), the bizarre shapes of the cysts seem to be the result of confluence (*black arrowheads*) (in focal illumination, such cysts appear greyish-white, cf. Fig. 1.6)

Comment

Accidental finding in the superior cornea in a 77-year-old woman after cataract surgery with superior incision; she had no symptoms relatable to the finding (see also Chap. 5, Case 3)

Signs of Pathology

Lines (1)

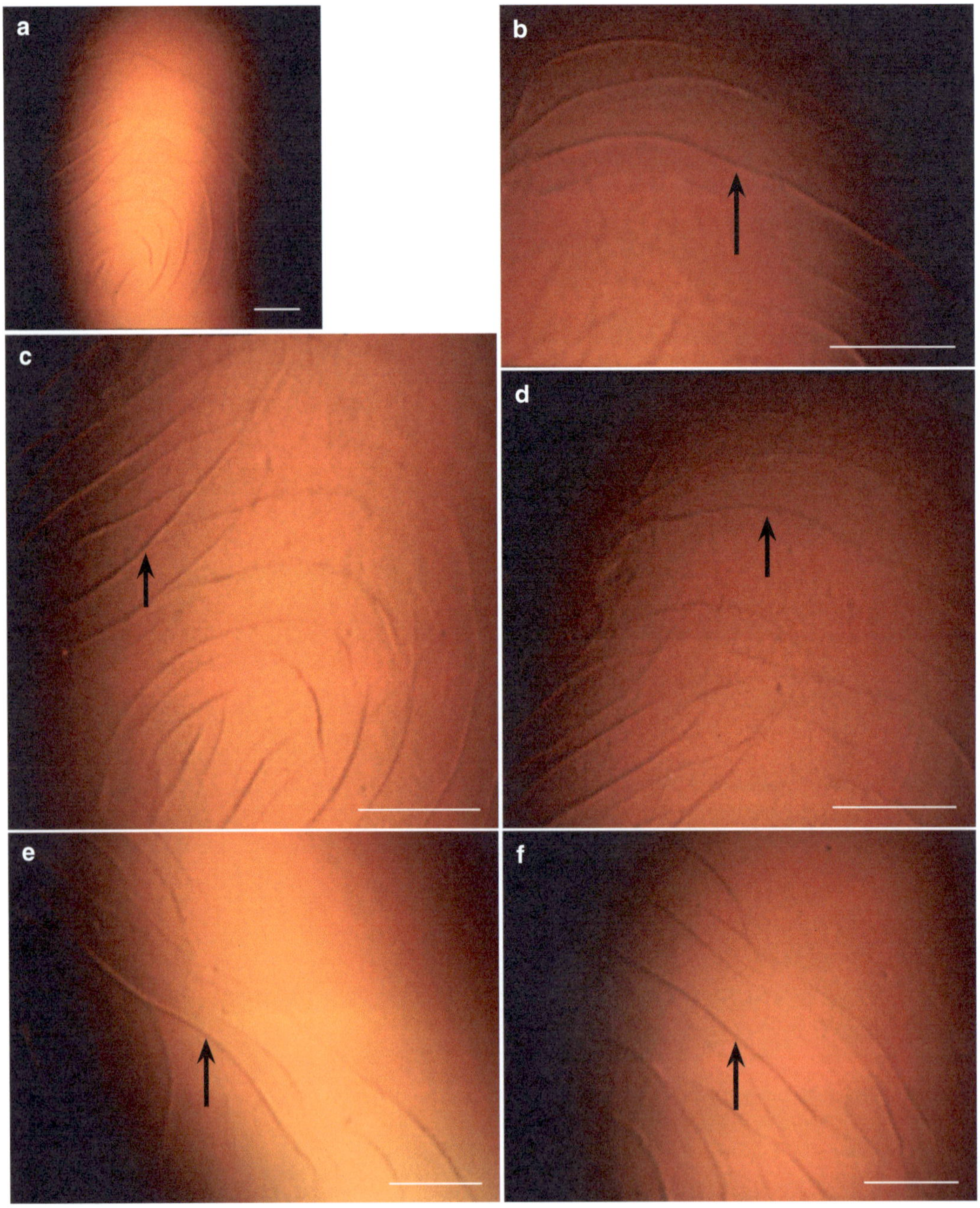

Fig. 1.8 (a) Survey. In low magnification, the distribution of the fine lines gives an impression of a fingerprint. (b–f) show variously thick lines (*arrows*) suggestive of ridges. The same cornea showed also small cysts (not shown)

Lines (2)

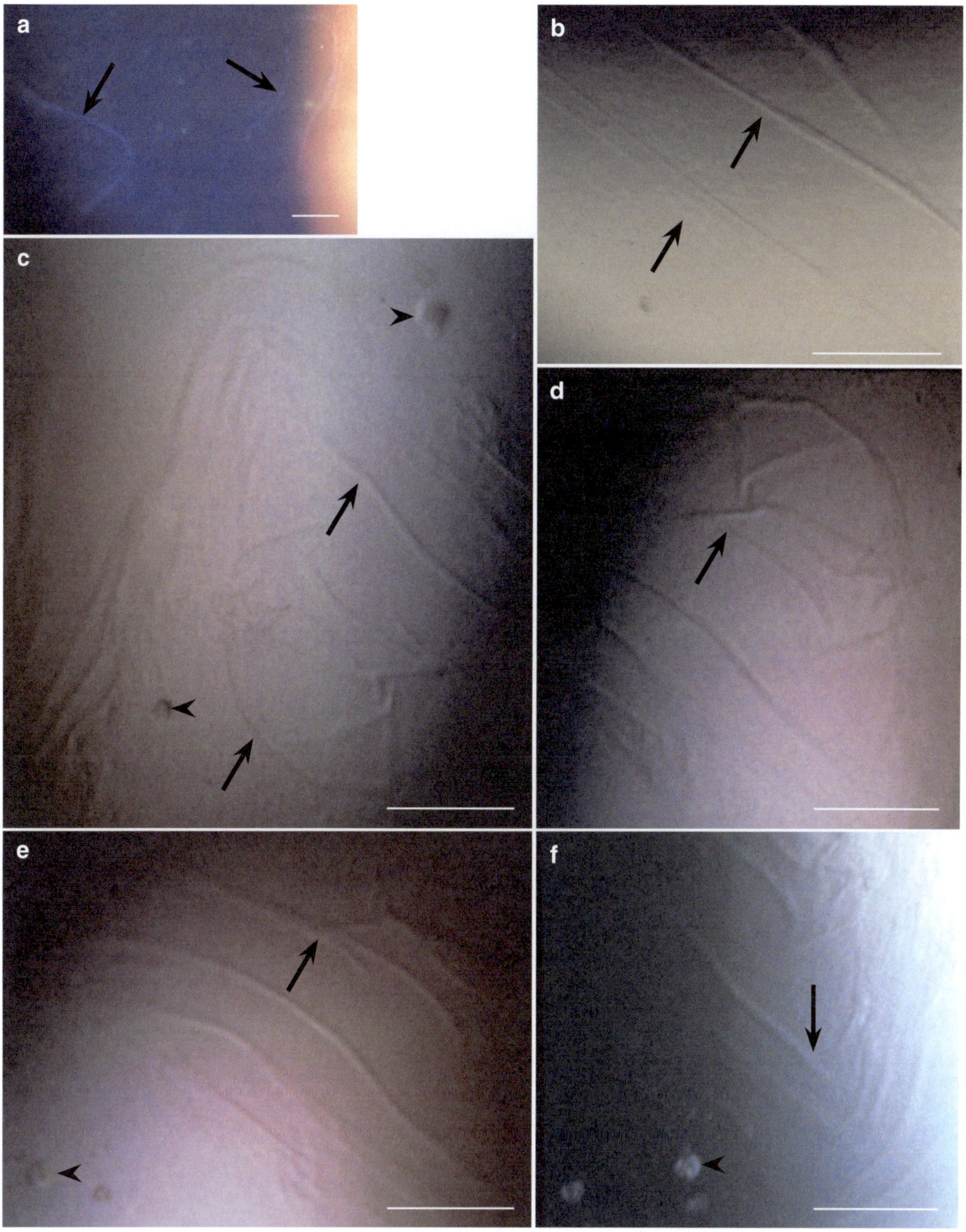

Fig. 1.9 (a–f) Variously shaped corneal lines (*arrows*). Those in (a) seem to represent thicker edges of epithelial membranes (cf. Figs. 1.10 and 1.11, overleaf). In (c, e and f) are additionally visible small cysts (*arrowheads*)

Signs of Pathology

Lines and 'Maps' (1)

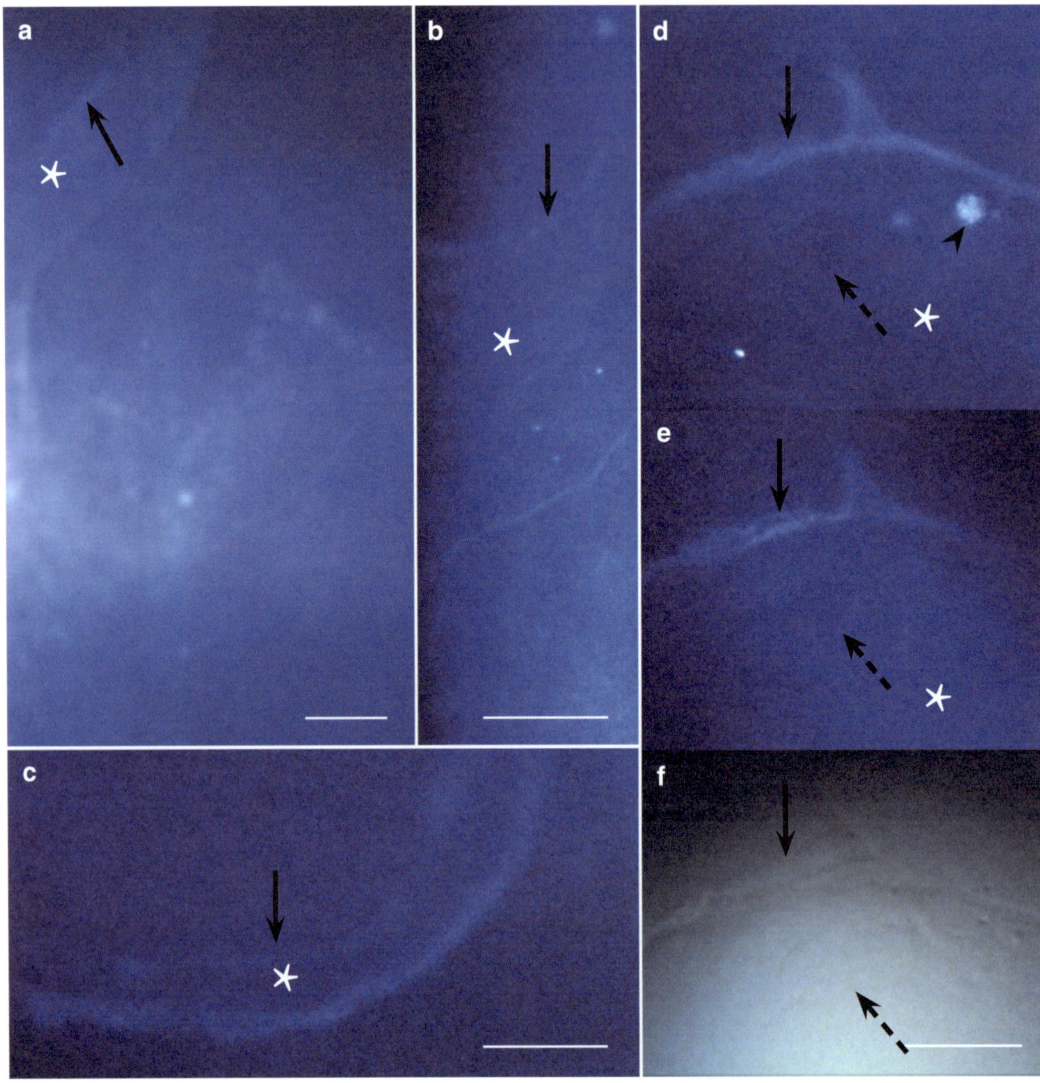

Fig. 1.10 (a–e) The edges of greyish membranes ('maps') (*asterisks*) often show brighter lines (*plain arrows*). In (d) a darker 'hole' (*dashed arrow*) in the membrane and a small light-reflecting cyst (*arrowhead*) are additionally visible. (e–f) show the same area 3 weeks later; the cyst has disappeared. In retroillumination (f), the light-reflecting line (*plain arrow*) appears thick. In (f), the *dashed arrow* indicates a faintly visible edge of the dark 'hole'. (The arrows in (d–f) are placed in corresponding locations; see also Chap. 2, Case 13)

Lines and 'Maps' (2)

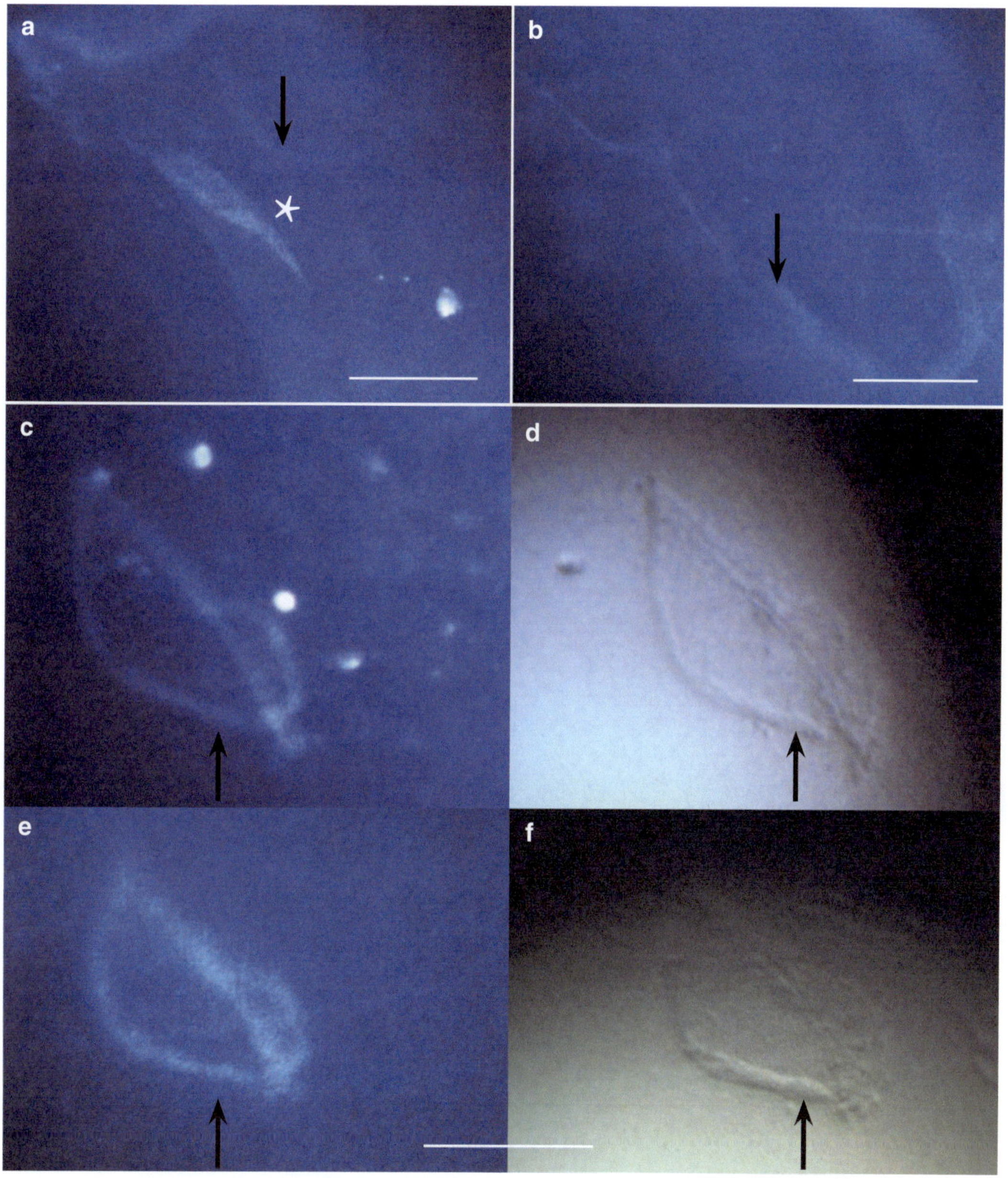

Fig. 1.11 (**a**, **b**) Two different areas of the same cornea. In (**a**) a greyish membrane (*asterisk*) and in (**a**, **b**) variously thick light-reflecting lines (*arrows*) are visible. (**c**, **d**) Light-reflecting thick lines (*arrows*) (**c**) in focal illumination and (**d**) in retroillumination. (**e**, **f**) show the same phenomena (*arrows*) in the same area 5 months later

Signs of Pathology

Blebs (1)

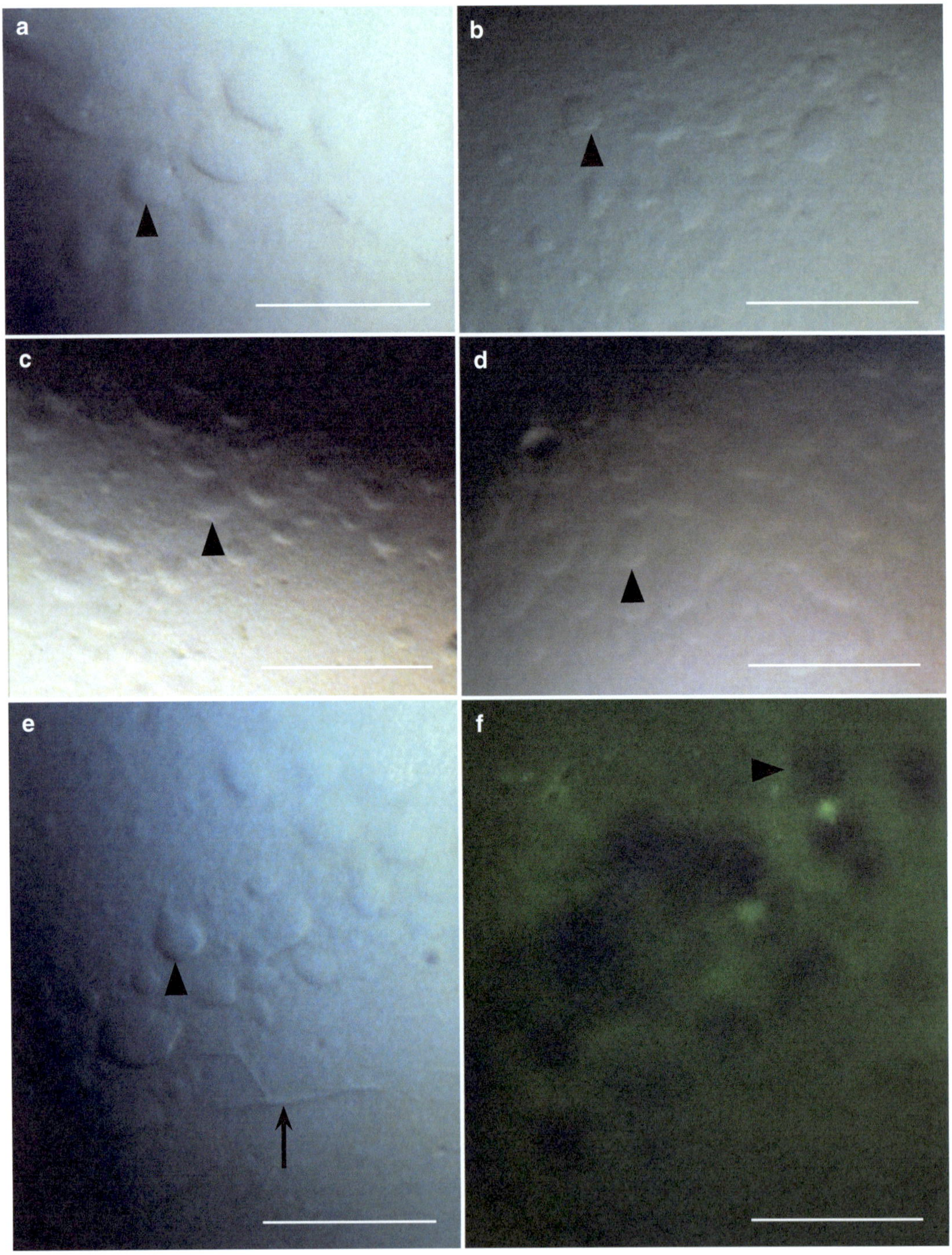

Fig. 1.12 (a–e) Variously large vesicle-like rounded structures (blebs, *arrowheads*) and (e) lines (*arrow*; cf. Fig. 2.2a). (f) Large blebs may occasionally cause surface elevations (*arrowhead*) visible as dark spots in the tear film stained green with fluorescein

Blebs (2)

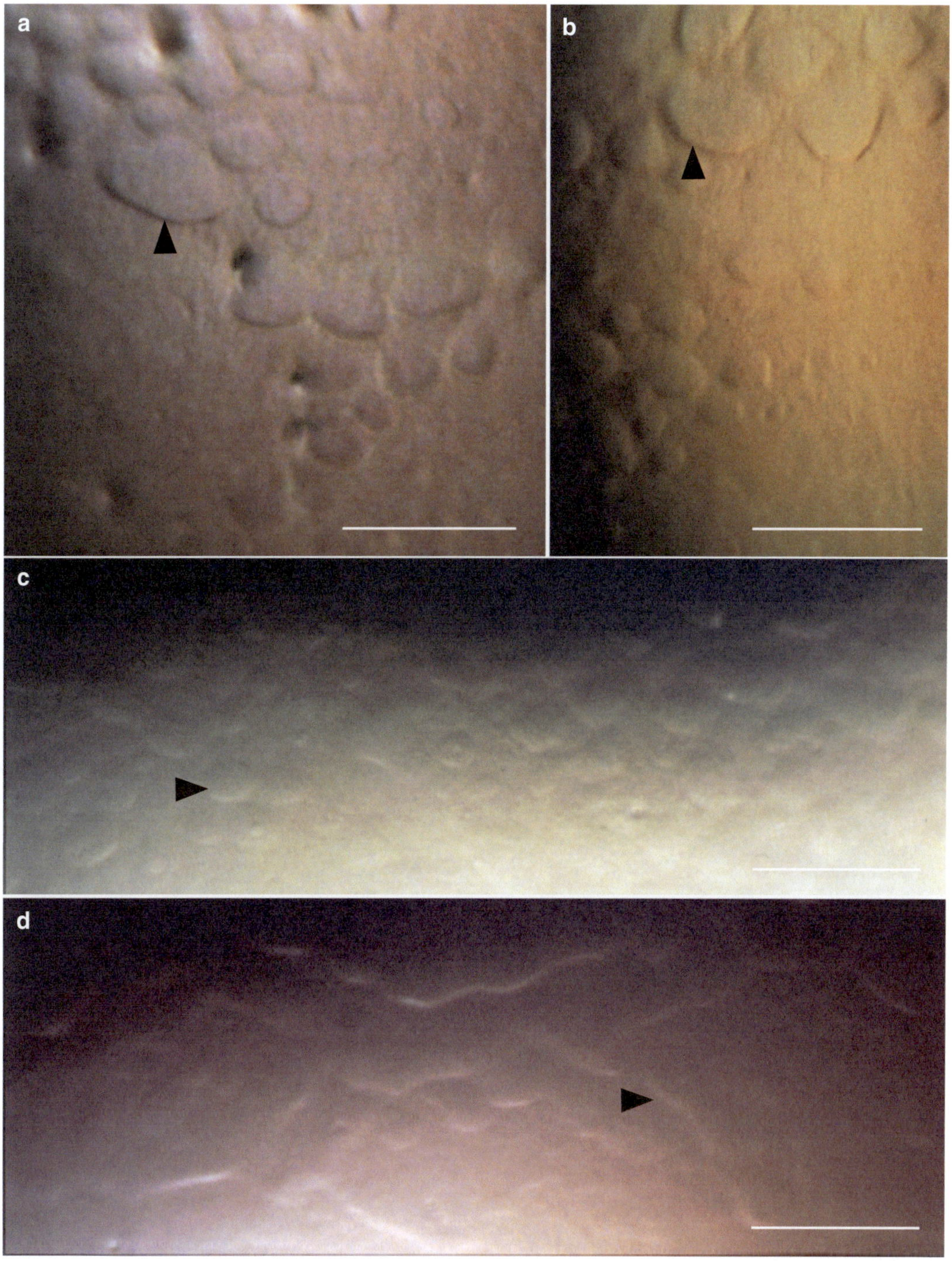

Fig. 1.13 (a–c) Variously large vesicle-like rounded structures (blebs, *arrowheads*). In (d) the blebs (*arrowhead*) seem partly confluent

Signs of Pathology

Various Cysts and Lines in Combination (1)

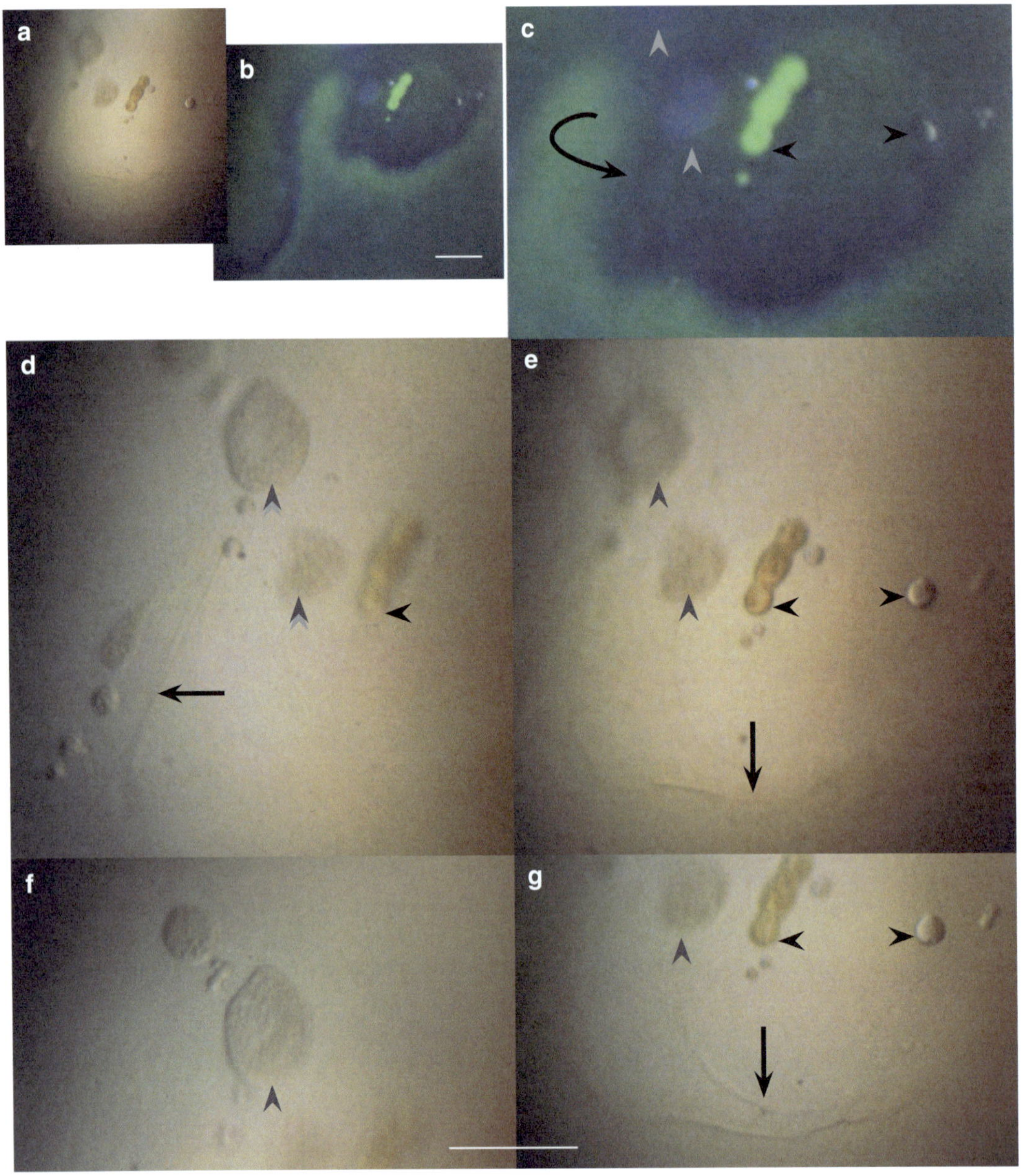

Fig. 1.14 (a, b) Survey of an area showing cysts, lines and surface elevations (a) before and (b) after the application of fluorescein. (c) In the green stained tear film, the area of surface elevation (*arrow*) appears dark; within this area, one of the opaque cysts (*grey arrowheads*) appears as a greyish not staining fleck (*lower grey arrowhead*), three rounded cysts in apposition stain brilliantly green (*left black arrowhead*) and another rounded cyst does not stain with the dye (*right black arrowhead*). (d–g) Several cysts, the opaque ones dense and granular (*grey arrowheads*), others round and apparently clear (*black arrowheads*). The *arrows* indicate lines, straight in (d) and bowed in (e, g) (the arrowheads are placed in corresponding locations)

Various Cysts and Lines in Combination (2)

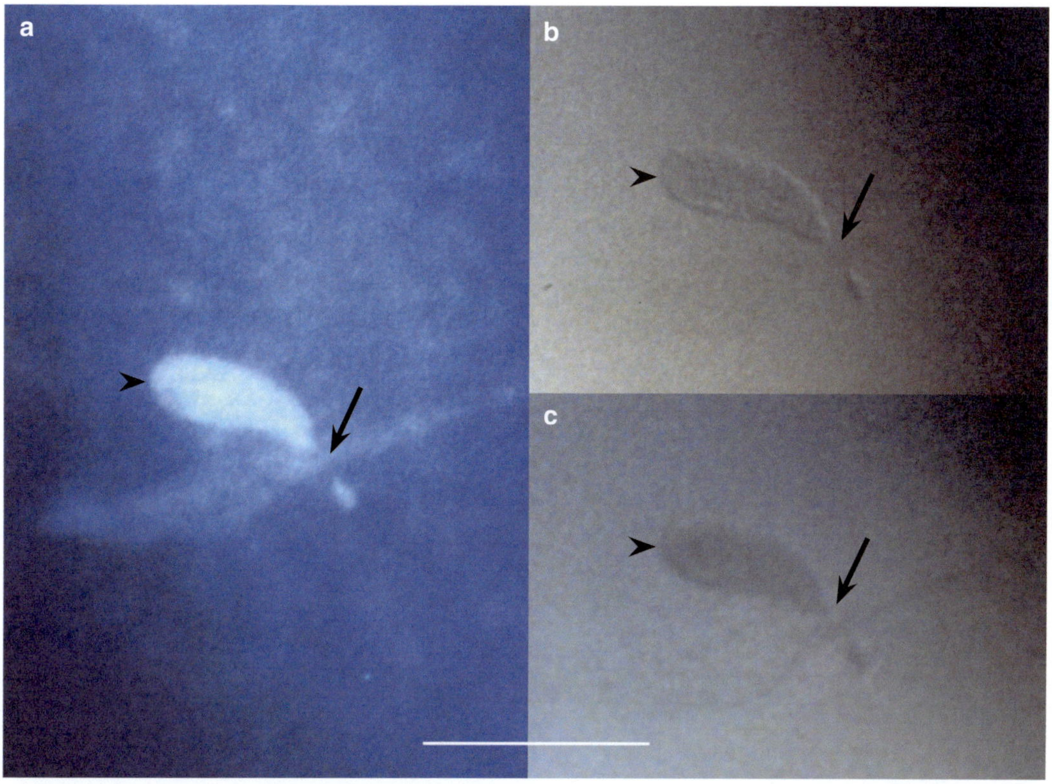

Fig. 1.15 (a) An opaque, strongly light-reflecting cyst (*arrowhead*) appearing (b, c) dense and (b) granular in retroillumination. The cyst seems situated below a membrane and trying to empty its contents by passing below the membrane's thicker edge (*arrows*) (cf. Figs. 2.47, 2.48 and 2.49)

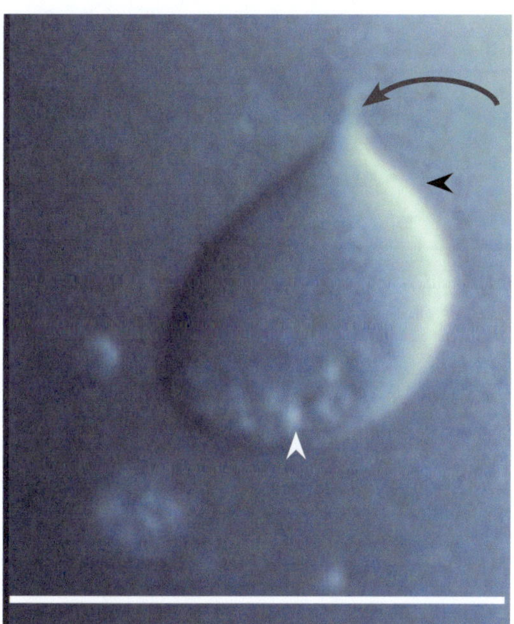

Fig. 1.16 A pear-shaped clear cyst (*black arrowhead*) the lower part of which contains rounded bodies/cell debris (*white arrowhead*). The cyst seems to have a surface connection (*arrow*) (cf. Fig. 2.52)

Signs of Repair

Epithelial Sheets (1)

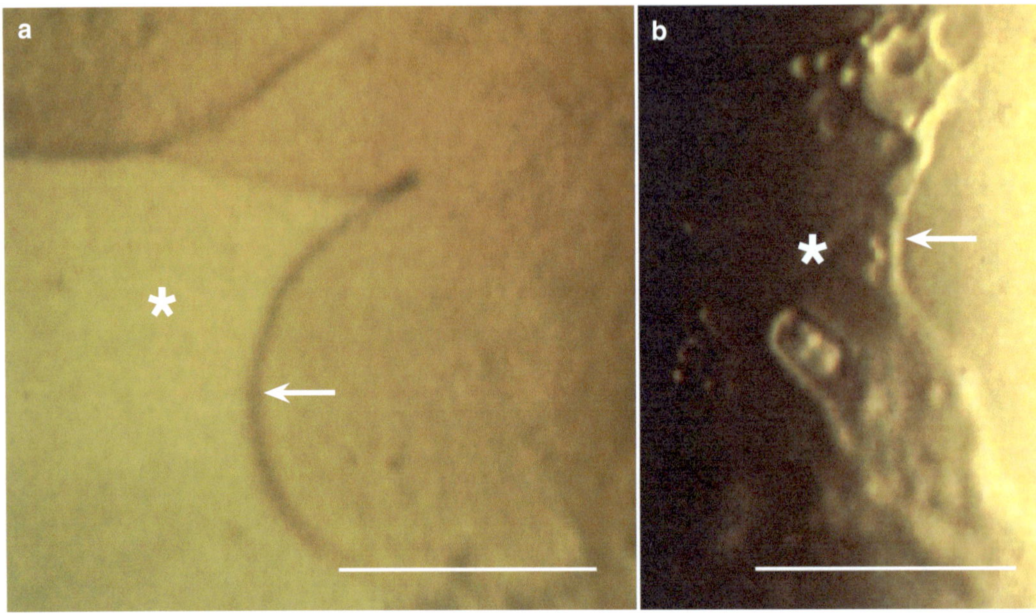

Fig. 1.17 (a) A curved tip (*arrow*) of an epithelial sheet pointing to a seemingly smooth surface (*asterisk*). (b) A similar tip (*arrow*) pointing to a deranged area (*asterisk*)

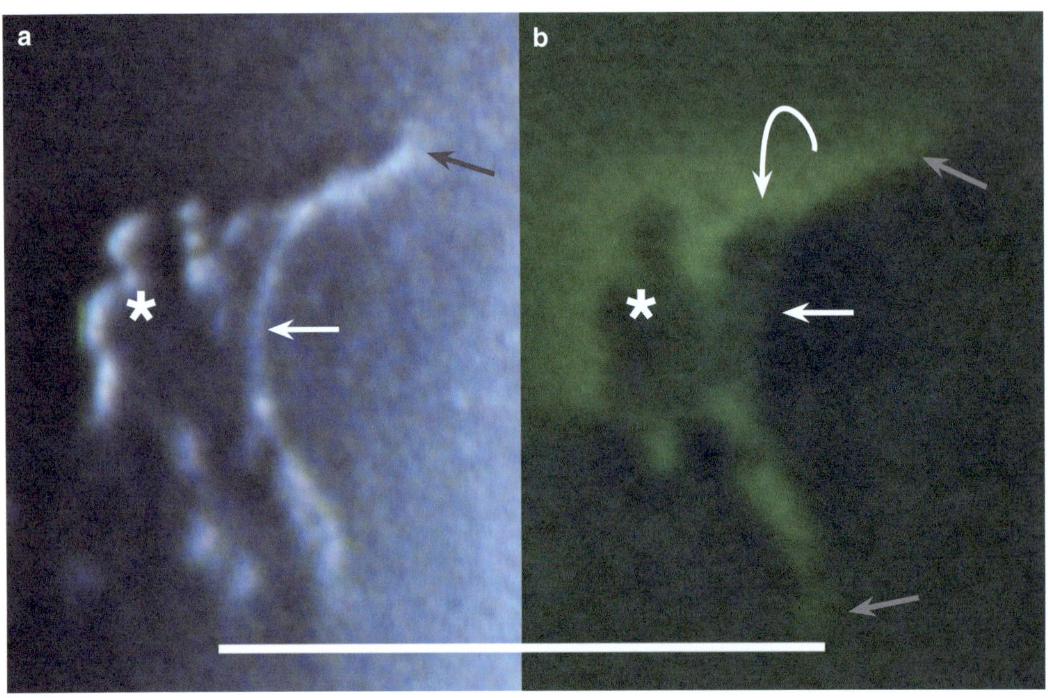

Fig. 1.18 (a) A small curved tip (*white arrows*) of an epithelial sheet. The *grey arrows* indicate areas of its merging with the surroundings. (b) In front of the tip, the green fluorescein pooling (*bowed arrow*) indicates a relative surface depression in which rounded structures (*asterisks*) are protruding (dark) (the markers are placed in corresponding locations)

Epithelial Sheets (2)

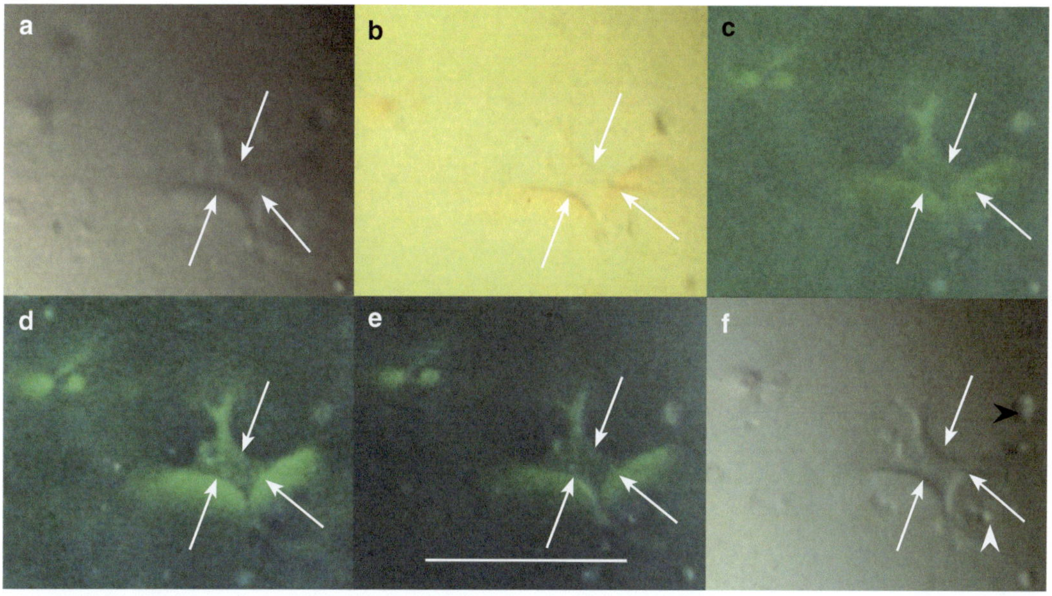

Fig. 1.19 Converging curved tips (*arrows*) of epithelial sheets. (a) Before staining. (b–e) Fluorescein pooling appears (b) yellow in retroillumination and (c–e) green in focal illumination. There is no diffusion of the dye into the surroundings. (f) The surroundings show rounded bodies (*white arrowhead*) and cystic changes (*black arrowhead*). (The photographs are shown in the order as taken. The arrows are placed in corresponding locations.) Compare Figs. 1.2, 1.20 and 1.21

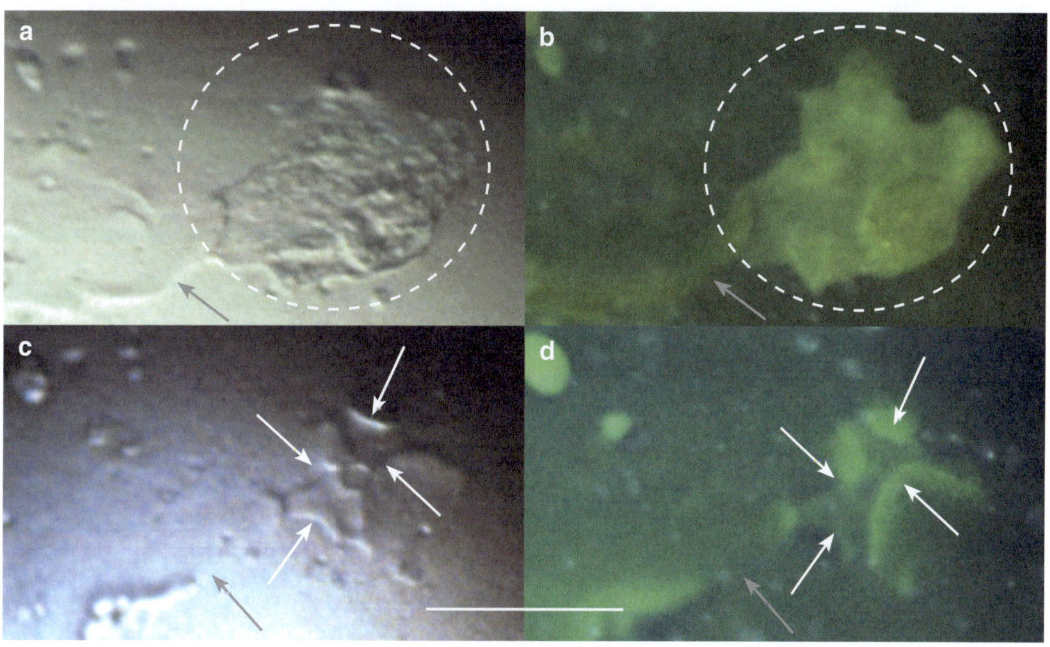

Fig. 1.20 (a, b) The area in *frame* indicates a part of a large lesion ((a) before and (b) after fluorescein staining) and the *arrows* a slightly curved edge. (c, d) 24 h later, the same area shows small converging tips of epithelial sheets (*white arrows*) and fluorescein pooling around and between them. The edge (*grey arrows*) has merged with its surroundings. (The grey arrows are placed in corresponding locations.) Compare Chap. 2, Case 9

Signs of Repair

Epithelial Sheets (3)

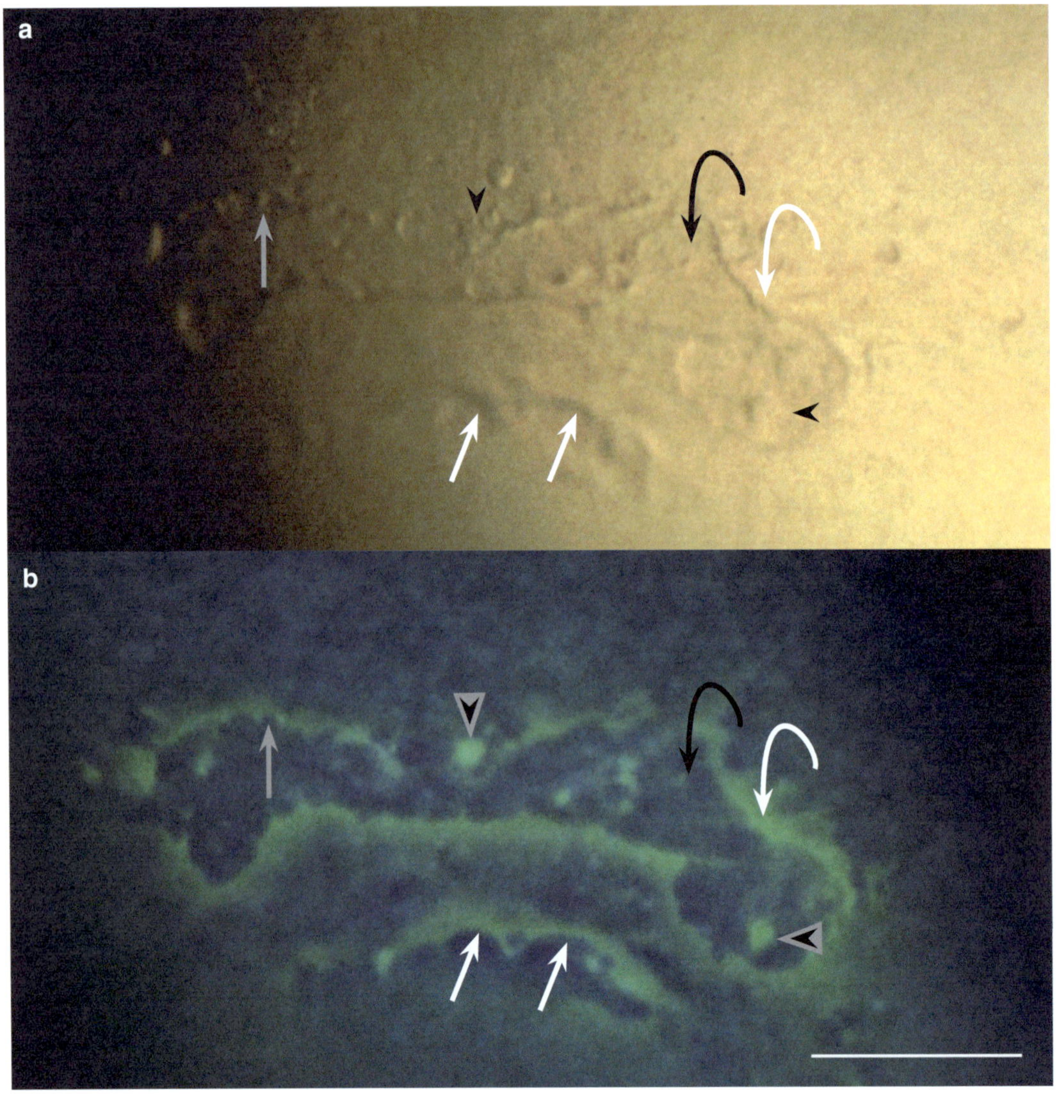

Fig. 1.21 A deranged area captured (a) before and (b) after the application of fluorescein. It shows cystic spaces (*arrowheads*), relative surface elevations (*black bowed arrows*) and surface depressions (*white bowed arrows*) in which fluorescein is pooling. The *straight white arrows* indicate curved lines implying smooth tips of epithelial sheets and the *grey arrows* irregularities at the upper edge of the area. (The markers are placed in corresponding locations.) Compare Fig. 1.22, opposite page

Epithelial Sheets (4)

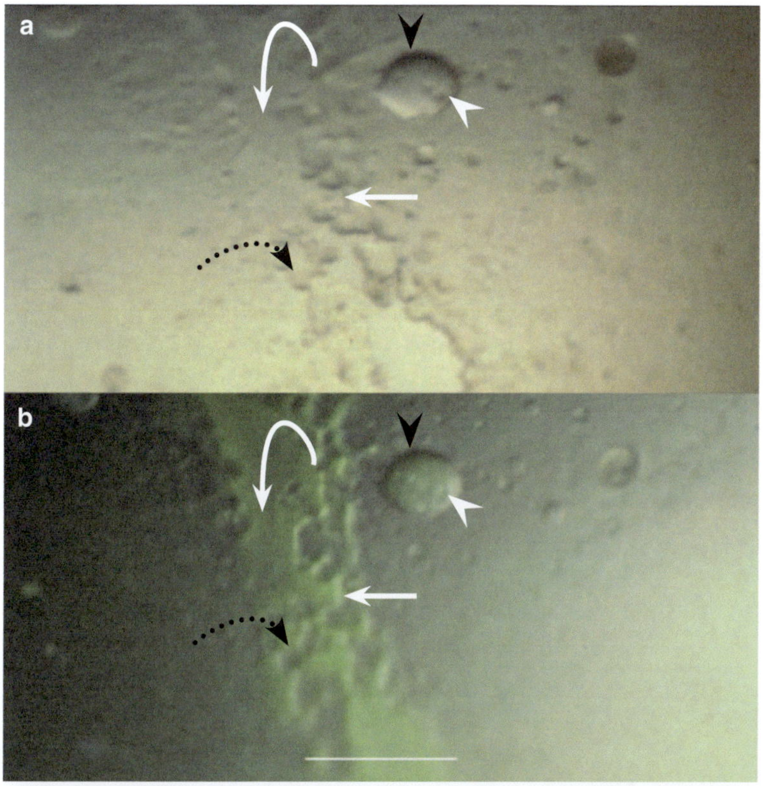

Fig. 1.22 A deranged area (**a**) before and (**b**) after the application of fluorescein. It shows a depression (*white bowed arrows*) in which fluorescein is pooling. The area contains some rounded elements (*black arrows*) and is surrounded by irregular borders (*straight arrows*). The surrounding epithelium shows cysts (*black arrowheads*), some containing rounded bodies (*white arrowheads*). (The markers are placed in corresponding locations.) Compare Figs. 1.28 and 2.42

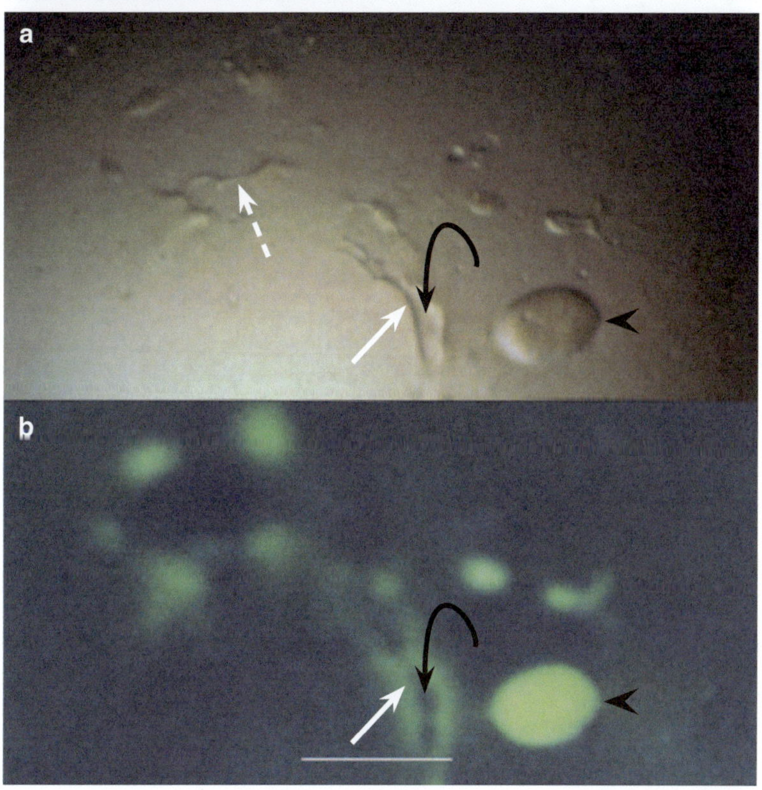

Fig. 1.23 (**a**) Before and (**b**) after the application of fluorescein. In this area a curved line is visible (*plain arrow*) implying a tip of an epithelial sheet; the structure in front of it (*bowed arrow*) is (**b**) elevated (dark). The *dashed arrow* in (**a**) indicates a short epithelial line (cf. Fig. 1.24, overleaf) and the *arrowhead* one of the cysts that (**b**) stain green with fluorescein (the markers are placed in corresponding locations)

Signs of Repair

Epithelial Lines and Fusiform Expansions (1)

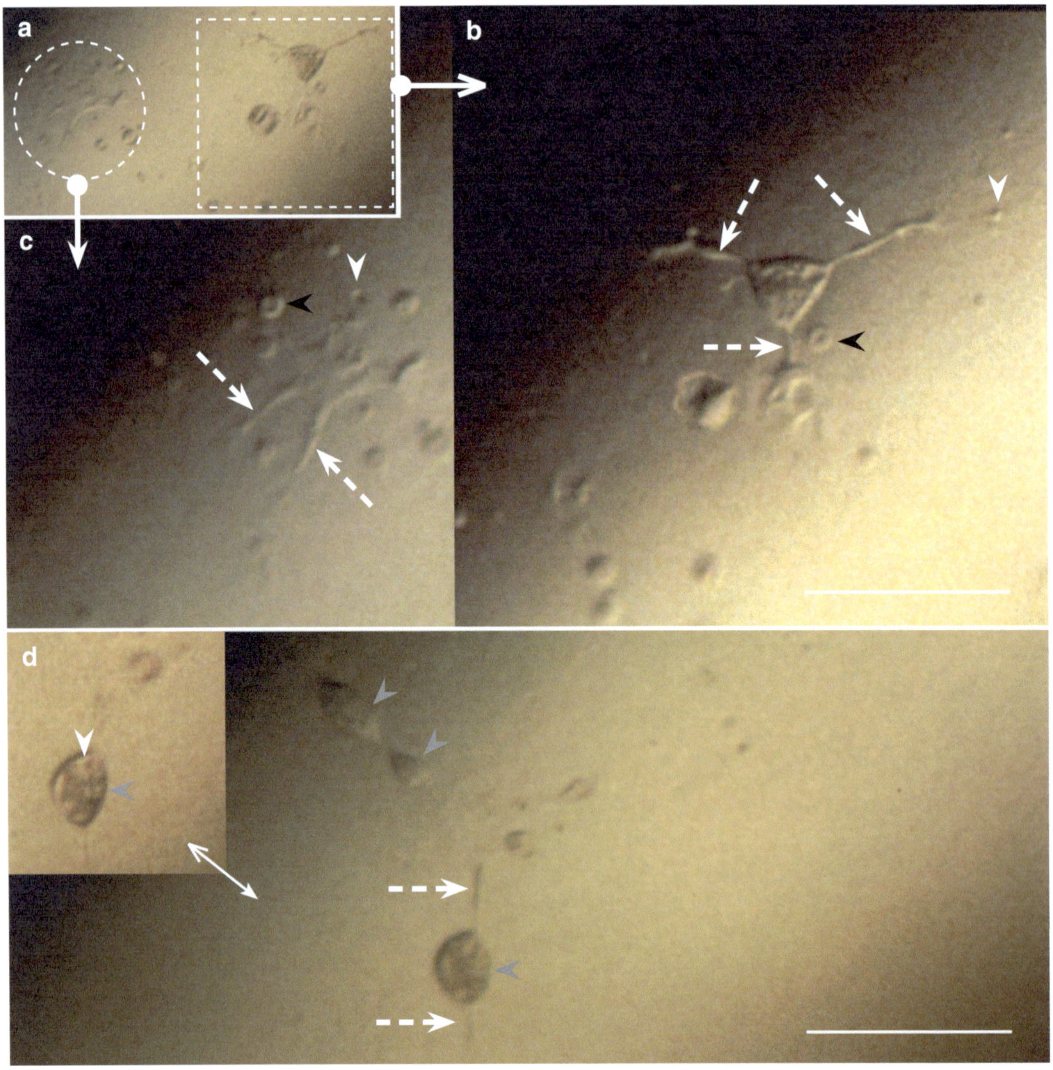

Fig. 1.24 (a) Survey. (b) The area in *rectangular frame* in (a) shows epithelial lines (*arrows*), cysts (*black arrowhead*) and rounded bodies (*white arrowhead*); such bodies are present also within the triangular space between the lines. (c) Also the area in *circular frame* in (a) shows epithelial lines (*arrows*), cysts (*black arrowheads*) and rounded bodies (*white arrowheads*). (d) In a different area of the same cornea are visible epithelial lines (*arrows*) and fusiform expansions (*grey arrowheads*) and in the *inset* rounded bodies (*white arrowhead*) within the largest one (*grey arrowhead*)

Fig. 1.25 (a–c) Survey of a deranged area captured in various illumination modes (a, b) before and (c) after the application of fluorescein. (a, b) show many light-reflecting structures and epithelial lines (*arrows*), which (c) after the application of fluorescein appear green. (d, e) The areas in *frames* in (a–c) show cysts (*black arrowheads*), epithelial lines (*dashed arrows*), fusiform expansions (*grey arrowheads*) containing rounded bodies ((d), *white arrowhead*) and (e) a line (*plain arrow*) implying a curved edge of an epithelial sheet. These structures are visualised with fluorescein (cf. (c)). (The dashed arrows are placed in corresponding locations in all photographs.) Compare Figs. 2.31 and 2.44

Epithelial Lines and Fusiform Expansions (2)

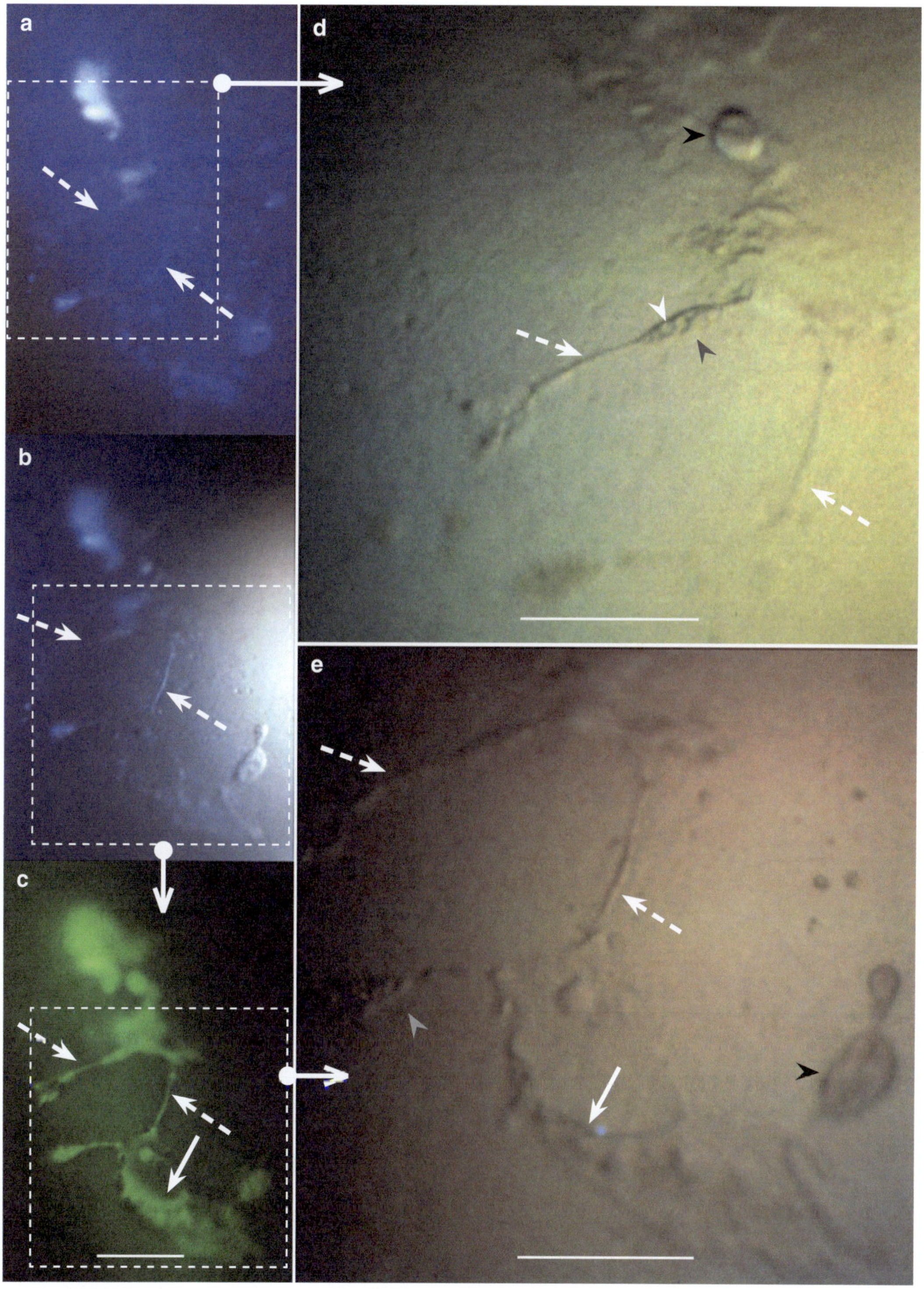

Signs of Repair

Epithelial Lines and Pear-Shaped Cysts

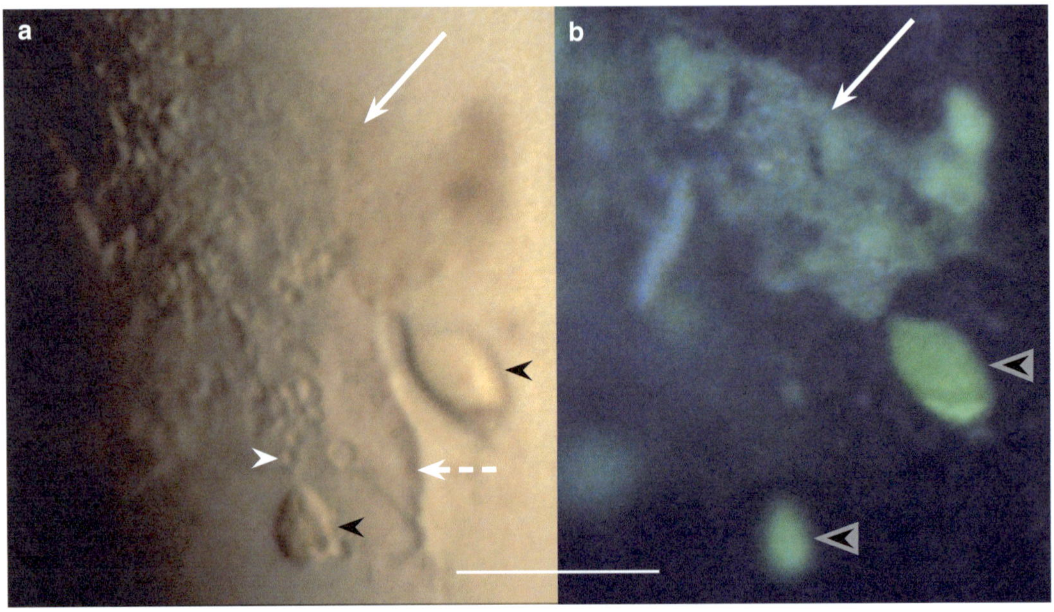

Fig. 1.26 (**a**) A part of a deranged area (*long arrow*) showing cysts (*black arrowheads*), rounded bodies (*white arrowhead*) and epithelial lines (*dashed arrow*). (**b**) Fluorescein has disappeared from the tear film; the green stained fluid has been retained within the upper part of the area (*long arrow*) and within the cysts (*arrowheads*) (the markers are placed in corresponding locations)

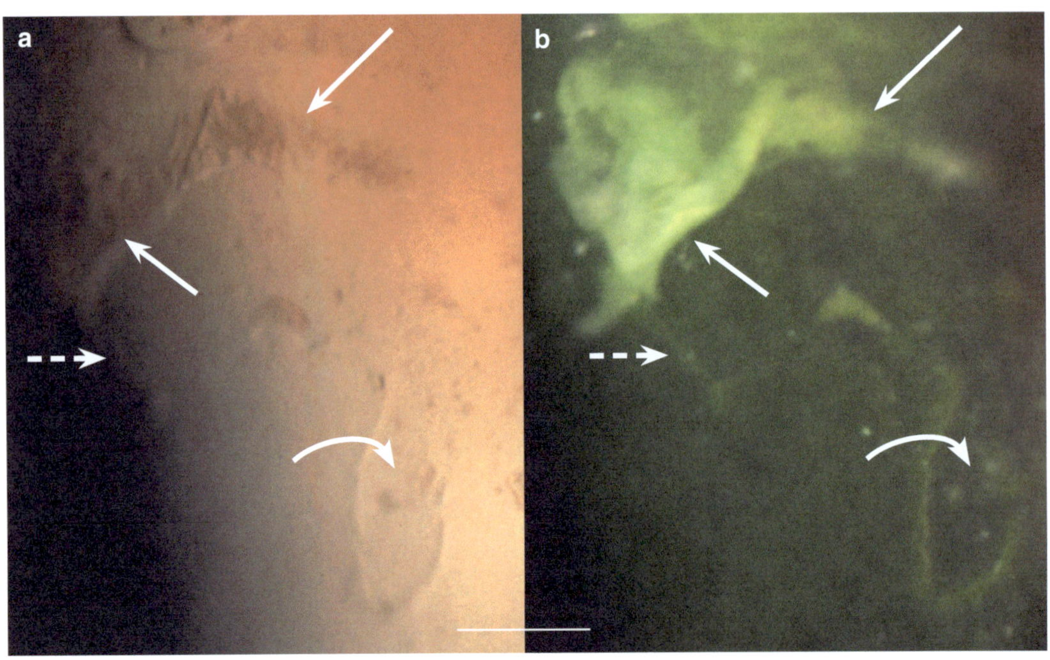

Fig. 1.27 (**a**) A deranged area the upper part of which (*long arrows*) is (**b**) light-reflecting. The *short plain arrows* indicate a curved line implying an edge of advancing epithelial sheet, the *dashed arrows* an epithelial line and the *bowed arrows* an area resembling a peninsula surrounded by a line in which (**b**) fluorescein is pooling (cf. Fig. 1.28, opposite page) (the markers are placed in corresponding locations)

Epithelial Lesions

Surface Elevations and Depressions

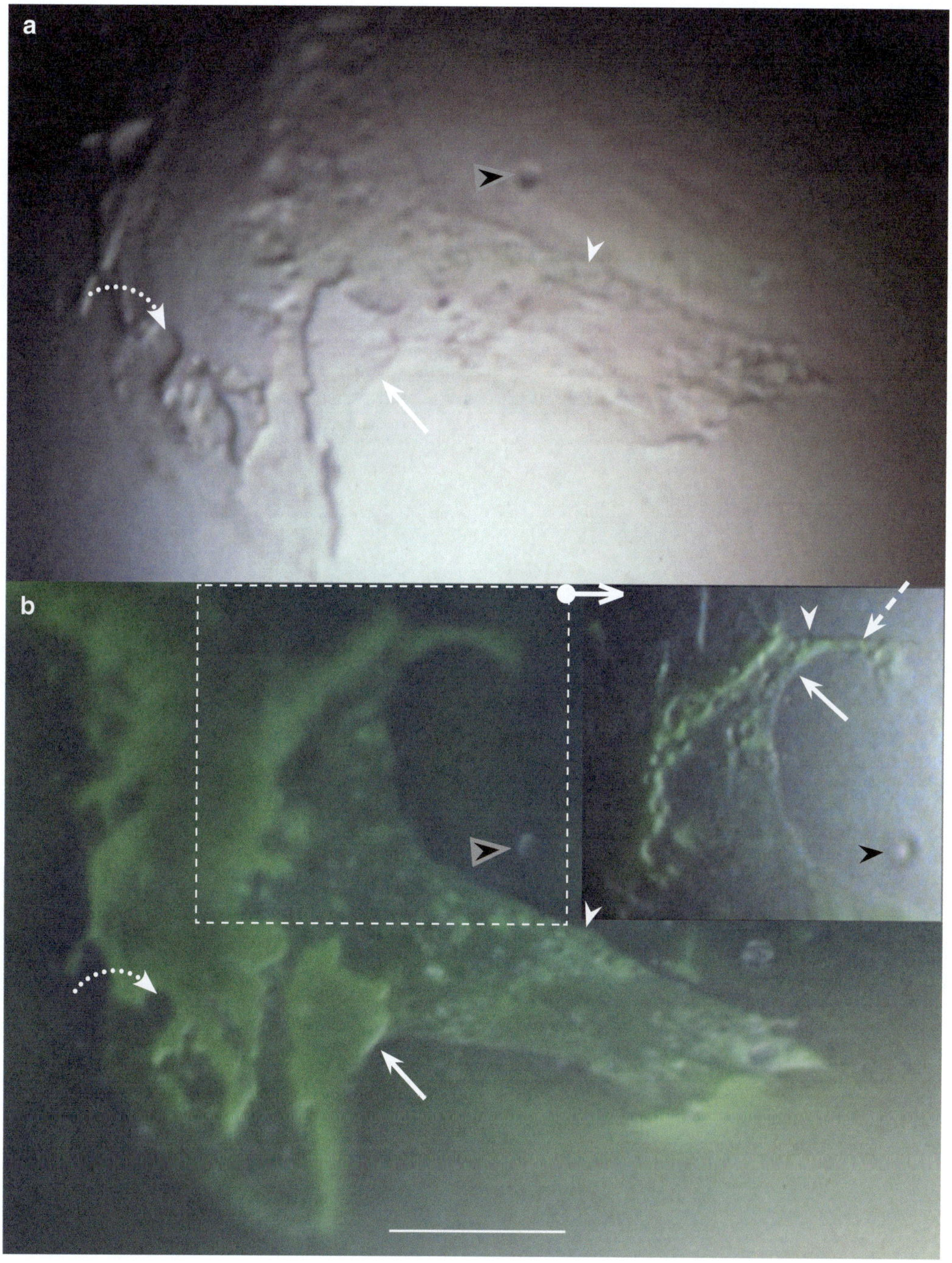

Fig. 1.28 (a) This area shows surface irregularities implying various levels in depth, in (b) visualised with fluorescein by various hues of green; relative elevations appear darker. The *plain arrows* indicate smooth, curved edges of epithelial sheets, the *bowed arrows* elevated structures at the irregular edge (cf. Fig. 1.22), the *black arrowheads* cysts and the *white arrowheads* rounded bodies within the lesion. (The markers are placed in corresponding locations.) *Inset* in (b) shows a cleft (*white dashed arrow*) between epithelial sheets; it contains rounded bodies (*white arrowhead*). The *black arrowhead* indicates the same cyst as in (a, b). See also Fig. 2.43

Epithelial Lesions

Epithelial Lines and Rounded Islands (1)

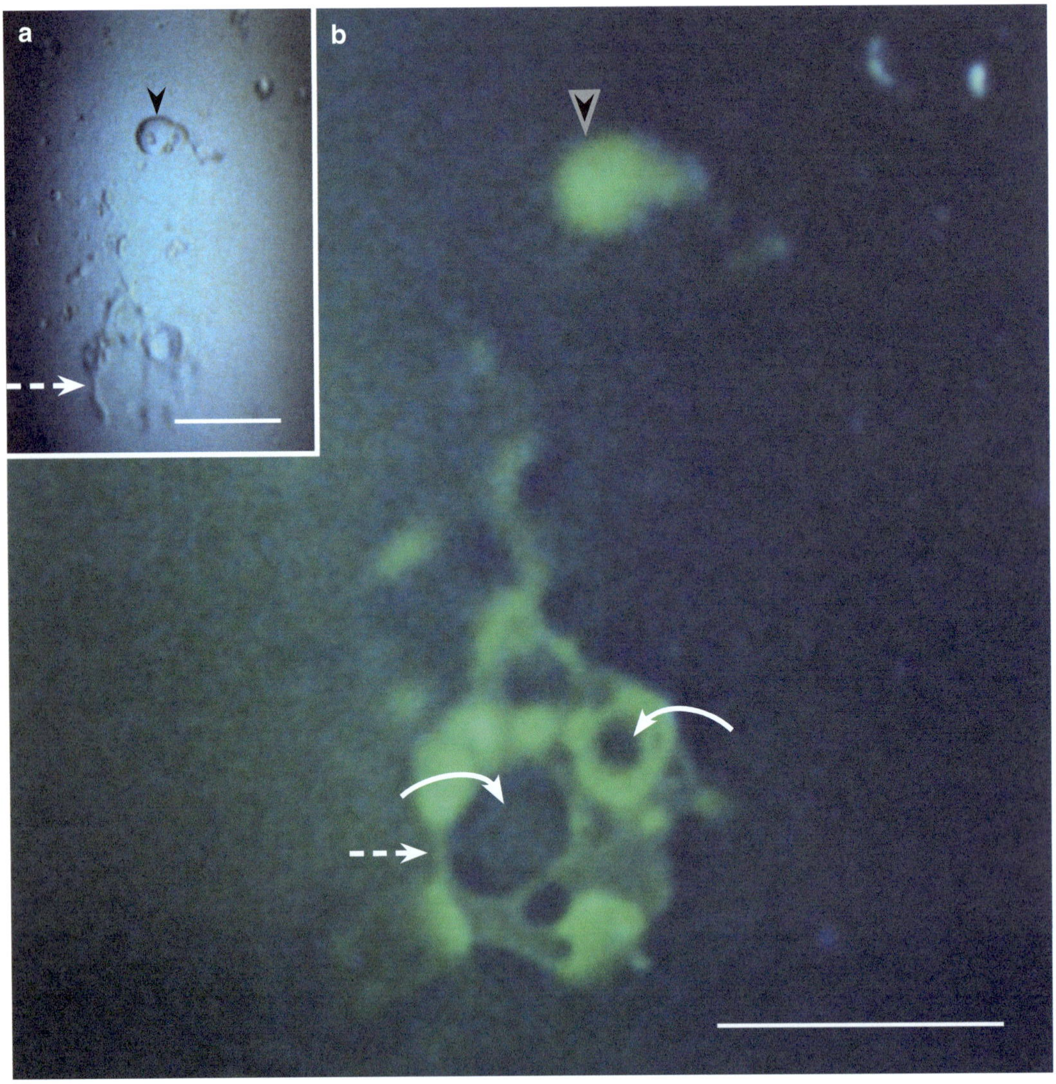

Fig. 1.29 (**a**) Survey of a deranged area showing epithelial lines (*arrow*) and cysts (*arrowhead*). (**b**) After the application of fluorescein becomes visible an area of well-delineated spaces showing various hues of green interspersed with dark rounded areas (islands, *bowed arrows*). In places, the less intensive green staining appears as lines (*dashed arrow*). The cyst in ((**a**), *arrowhead*) stains green with fluorescein ((**b**), *arrowhead*) (the arrowheads and the dashed arrows are placed in corresponding locations)

Epithelial Lines and Rounded Islands (2)

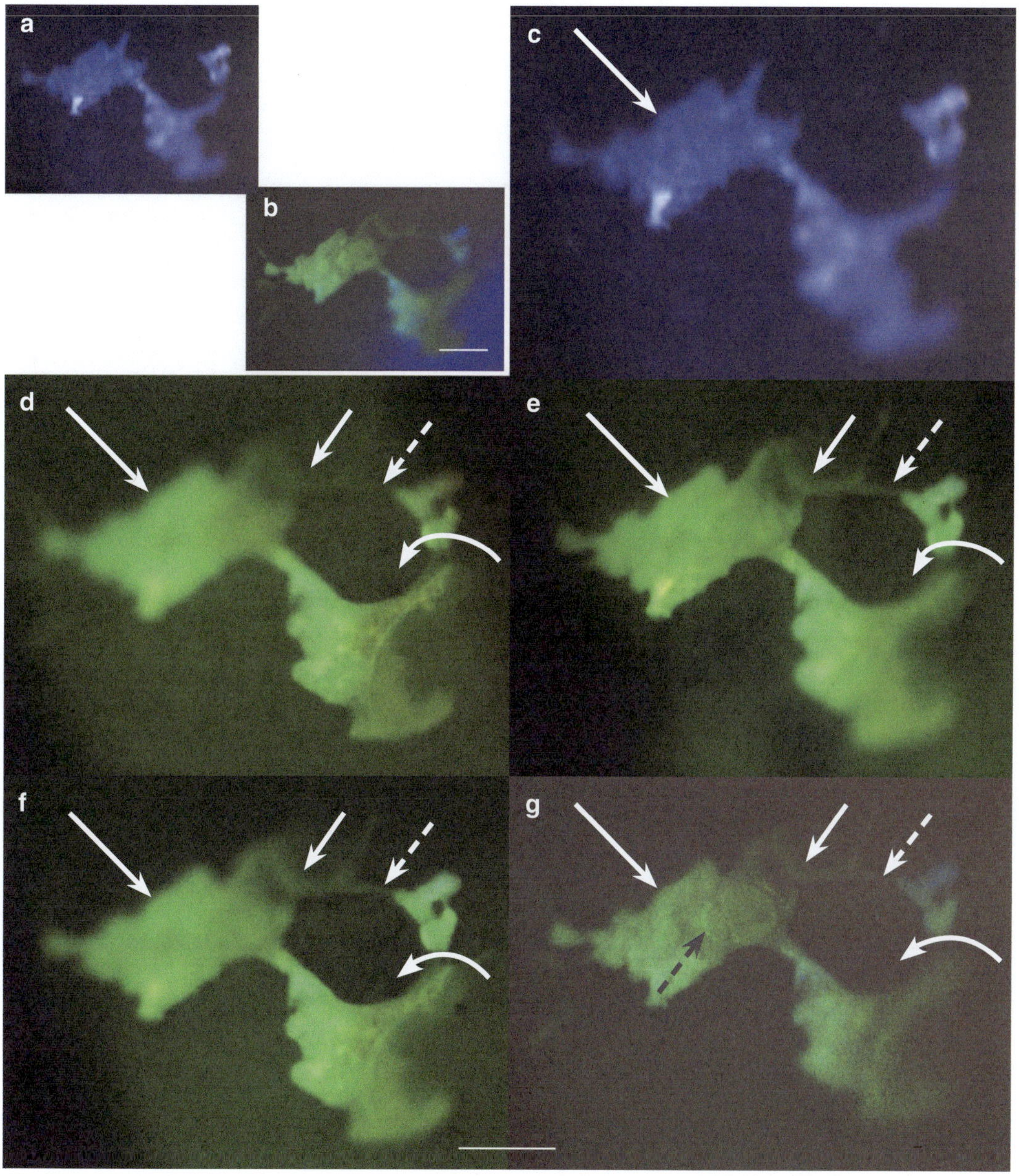

Fig. 1.30 (**a**, **b**) Survey of a lesion, (**a**) before and (**b**) after the application of fluorescein. (**c**) The lesion (*long arrow*) is strongly light-reflecting; its edges are wavy. (**d–g**) Sequential photographs capturing green fluorescein staining show a pattern corresponding to (**c**) and additional features such as a green line (*dashed arrows*) and another, curved green line implying a tip of an epithelial sheet (*short plain arrows*) completing delineation of a rounded island (*bowed arrow*) with slightly wavy edges. The *grey dashed arrow* in (**g**) indicates well-delineated oval area within the lesion (the markers are placed in corresponding locations; **b**, blue filter)

Epithelial Lesions

Epithelial Lines and Rounded Islands (3)

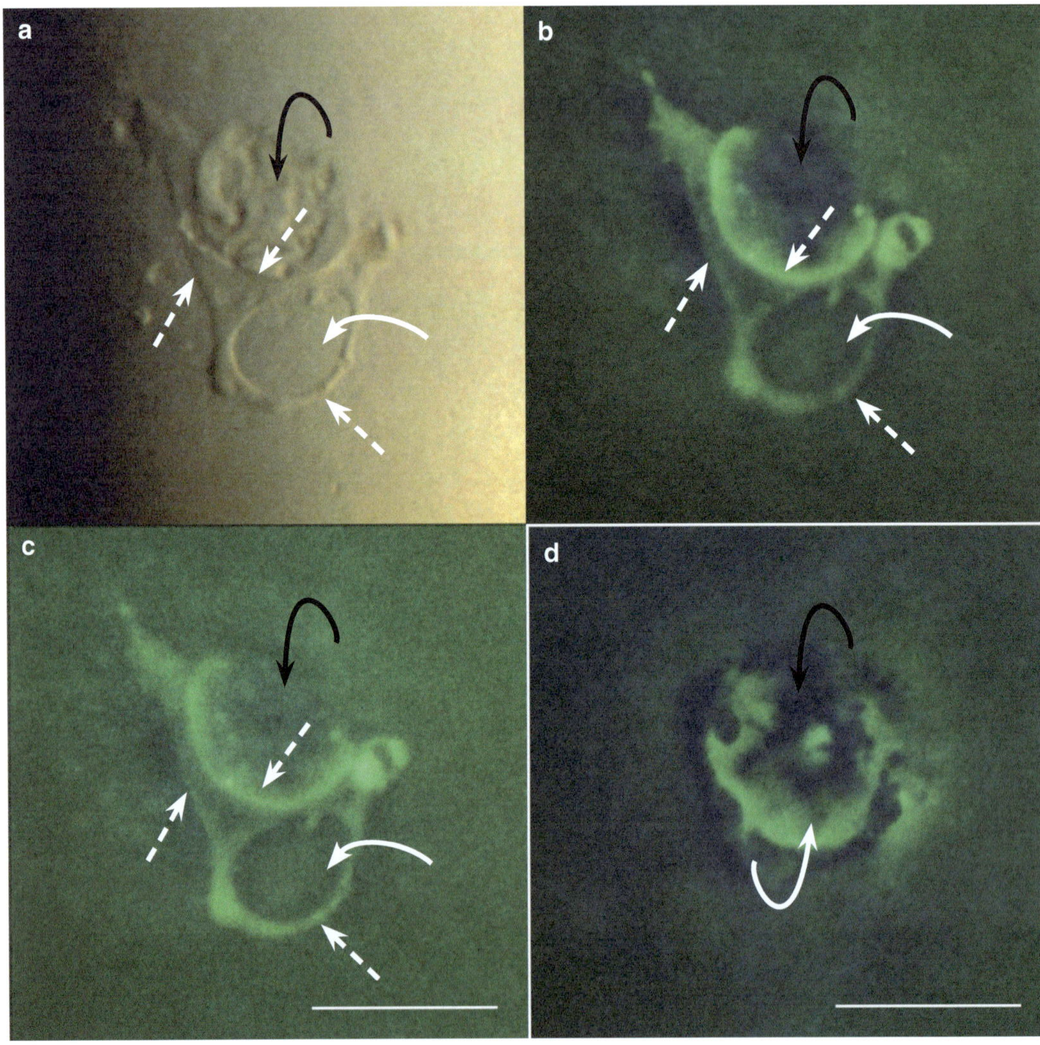

Fig. 1.31 (a) A lesion showing variously bowed epithelial lines (*straight arrows*), an area with irregular contents (*black bowed arrow*) and a smooth, well-delineated oval one (*white bowed arrow*). (b, c) With fluorescein it is visible that the irregular contents cause surface elevation (*black bowed arrow*) and that the dye is pooling in the lines (*straight arrows*); the oval area (*white bowed arrow*) appears as an elevated island. (The arrows are placed in corresponding locations.) (d) A different area of the same cornea showing fluorescein pooling (*white arrow*) and surface elevations (*black arrow*)

A Cracked Surface?

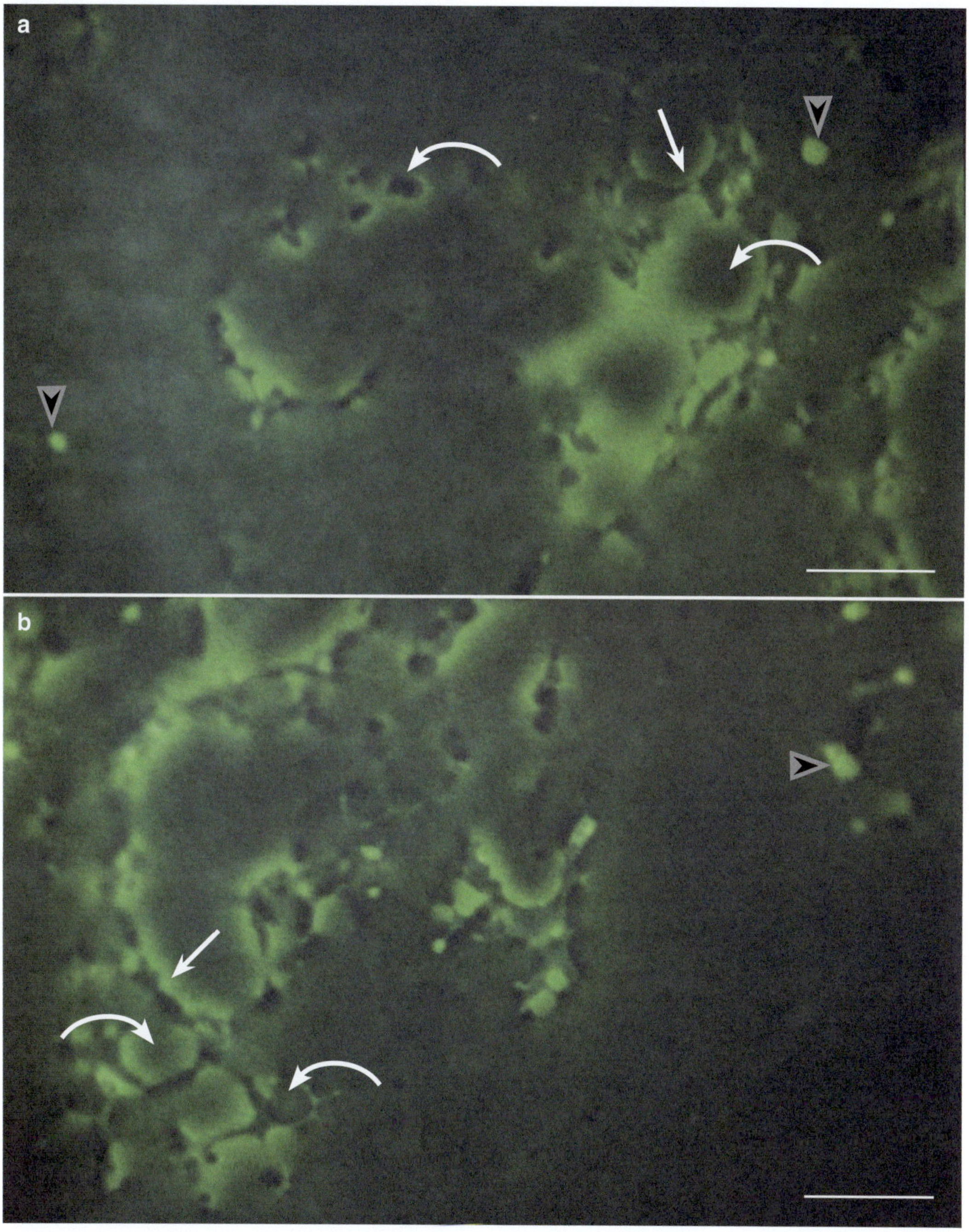

Fig. 1.32 (a, b) Two areas captured in the same cornea. By pooling, the green stained tear fluid reveals variously broad curved lines (*straight arrows*) in places surrounding larger and smaller rounded areas (*bowed arrows*); the interplay between brightly green and dark areas gives the impression of a cracked surface. The *arrowheads* indicate green stained cystic spaces

Complex Lesions (1/1)

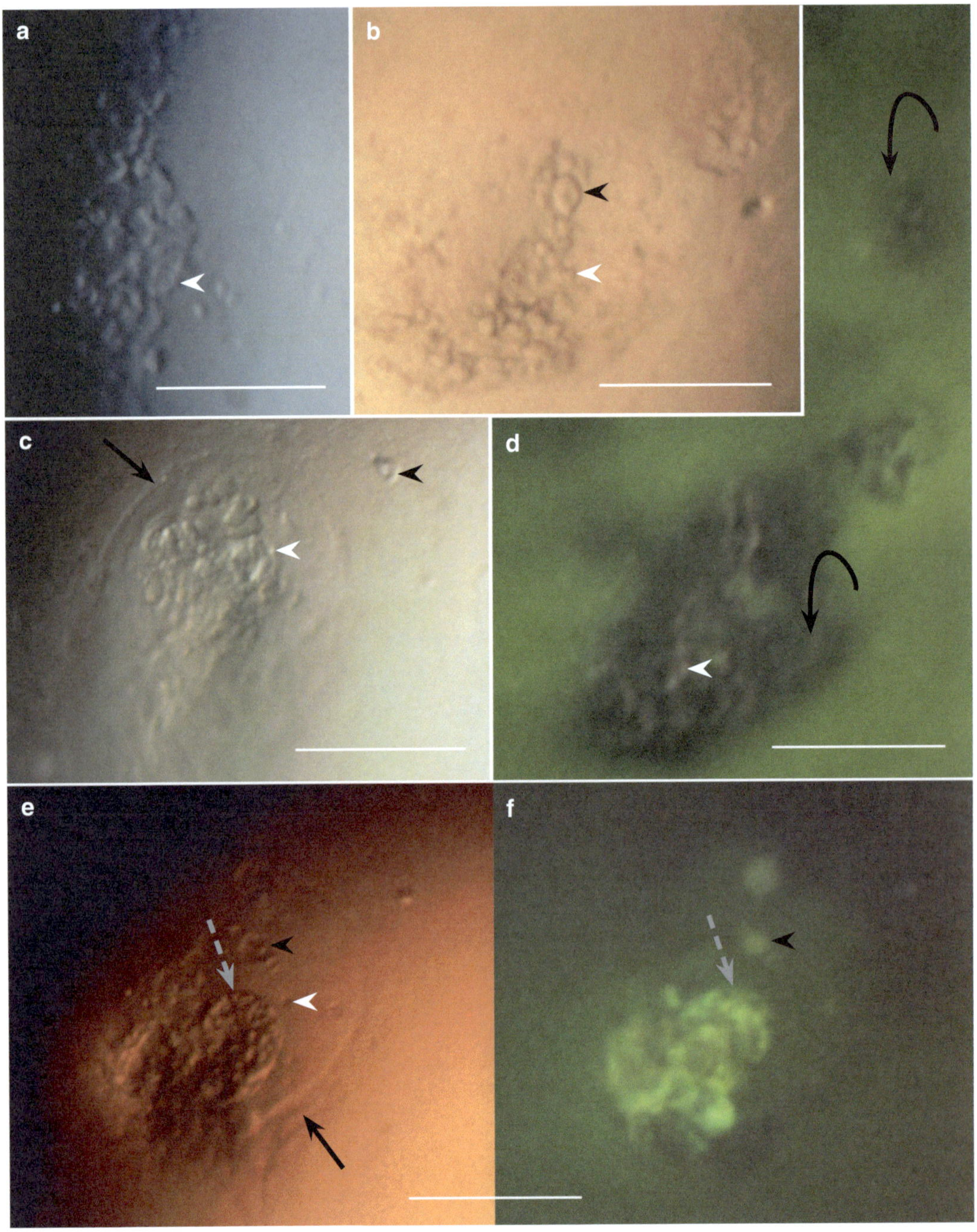

Fig. 1.33 In these lesions, the main component are heaped-up rounded bodies (a–e, *white arrowheads*); *black arrowheads* in (b, c, e and f) indicate small cysts. In (c, e) lines (*black straight arrows*) of fingerprint type are visible. (d) shows a larger and a smaller lesion, both causing elevations (dark in the green stained tear film, *arrows*) of an apparently undisrupted surface. (e, f) show another lesion, (e) before and (f) after fluorescein staining; the *dashed arrows* indicate an edge of a circular area within the lesion (cf. Fig. 1.34d, opposite page). Small cystic spaces (*black arrowheads*) stain green. None of the lesions shows a clearly defined edge (in e–f the dashed arrows and the black arrowheads are placed in corresponding locations)

Complex Lesions (1/2)

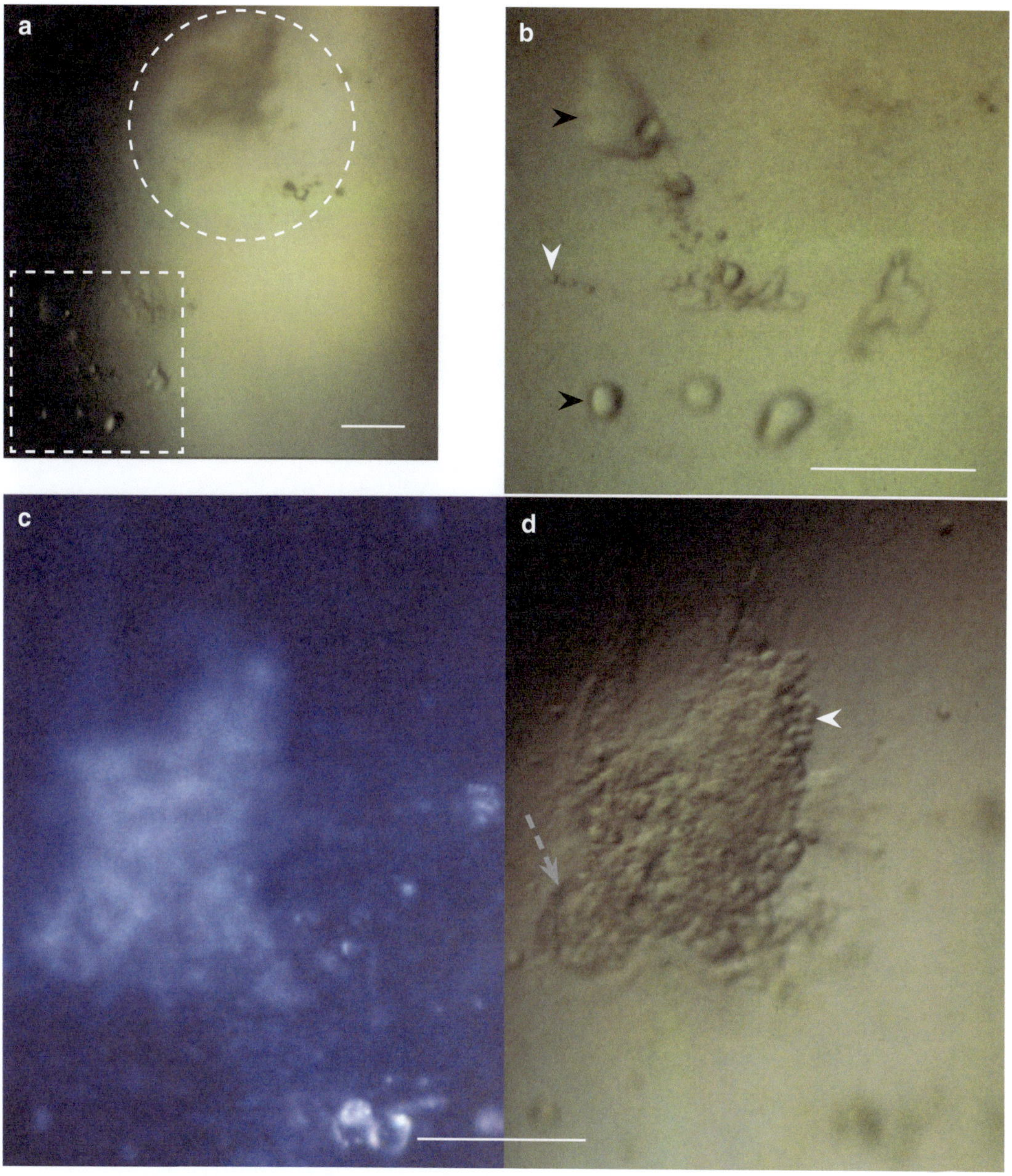

Fig. 1.34 (a) Survey. (b) In the area indicated by *rectangular frame* in (a) are visible cystic spaces (*black arrowheads*) and some rounded bodies (*white arrowhead*). (c–d) The area indicated by *circular frame* in (a) shows a lesion that is (c) strongly light-reflecting and (d) contains many rounded bodies (*arrowhead*; cf. Fig. 2.33); the *arrow* indicates a circular area within the lesion (cf. Fig. 1.33e, opposite page). The lesion partly lacks a sharply defined edge

Complex Lesions (2/1)

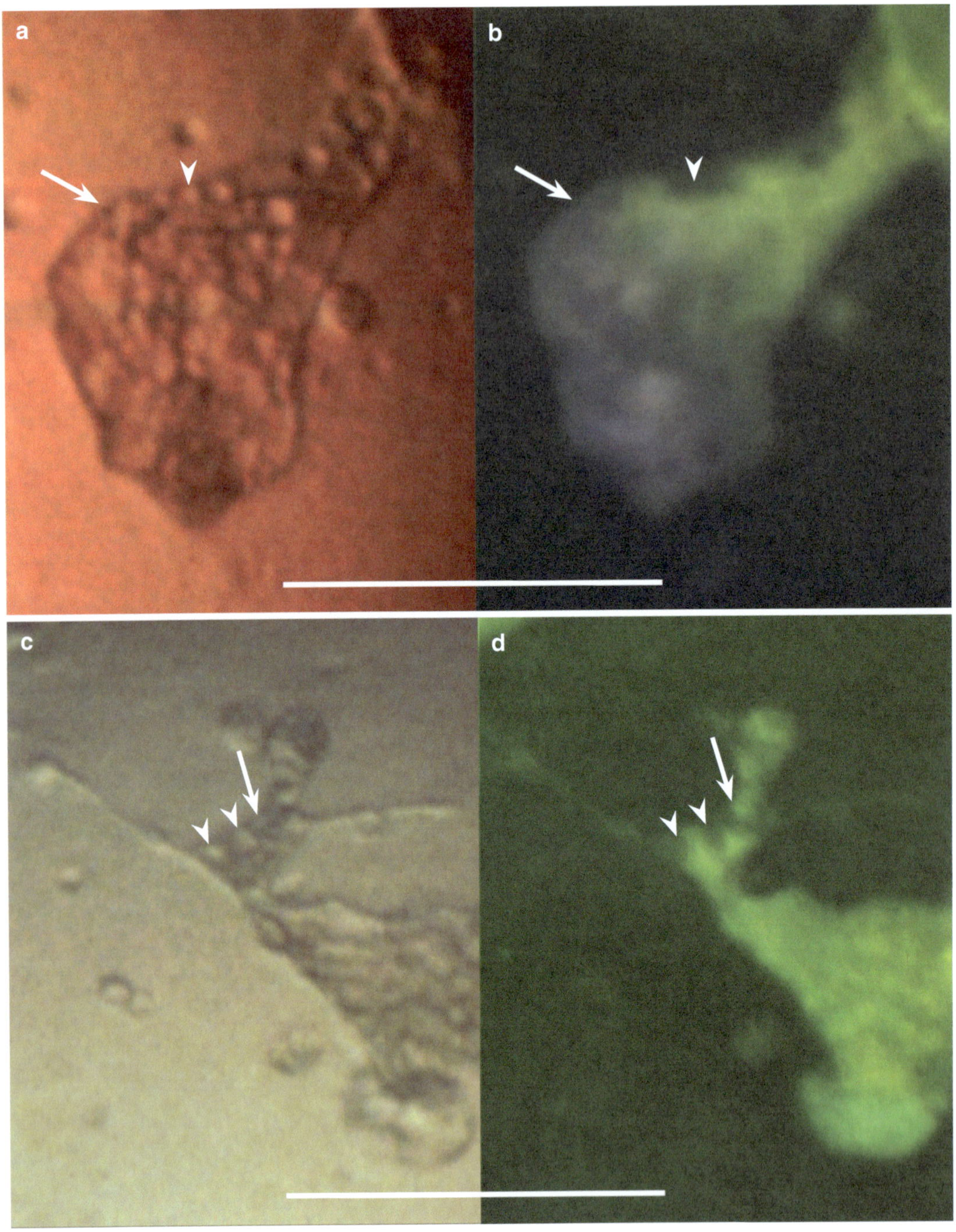

Fig. 1.35 (a, b) This part of a larger lesion (cf. Fig. 1.40) shows that (a) some rounded bodies (*arrowheads*) in front of the curved edges (*arrows*) of the lesion are (b) not light-reflecting. (c, d) The same phenomena captured in another lesion (cf. Fig. 1.42) (the markers are placed in corresponding locations in each pair of photographs)

Complex Lesions (2/2)

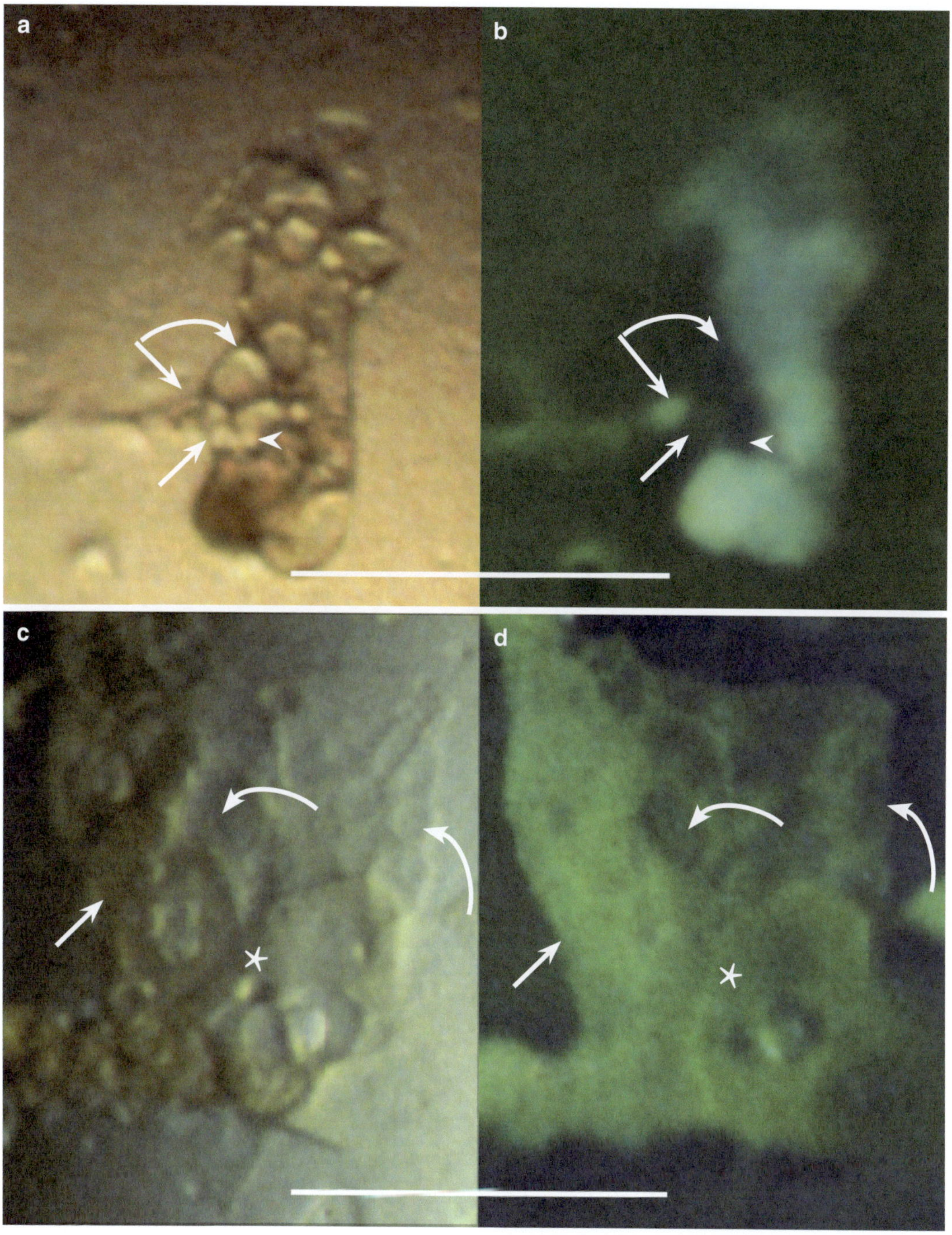

Fig. 1.36 (**a, b**) The *straight arrows* indicate edges of a part of a larger lesion (cf. Fig. 1.41) showing rounded bodies (*arrowhead*) and larger rounded structures (*bowed arrow*). In (**b**) it is visible that the structures are neither light-reflecting nor stain with fluorescein. (**c, d**) In this part of another larger lesion (cf. Figs. 2.13 and 2.14) rounded structures (*bowed arrows*) are visible which (**d**) appear elevated (dark) in the green stained tear fluid; the *plain arrows* indicate one edge of the lesion and the *asterisks* a relatively depressed oval area in which fluorescein is pooling (the markers are placed in corresponding locations in each pair of photographs)

Complex Lesions (2/3)

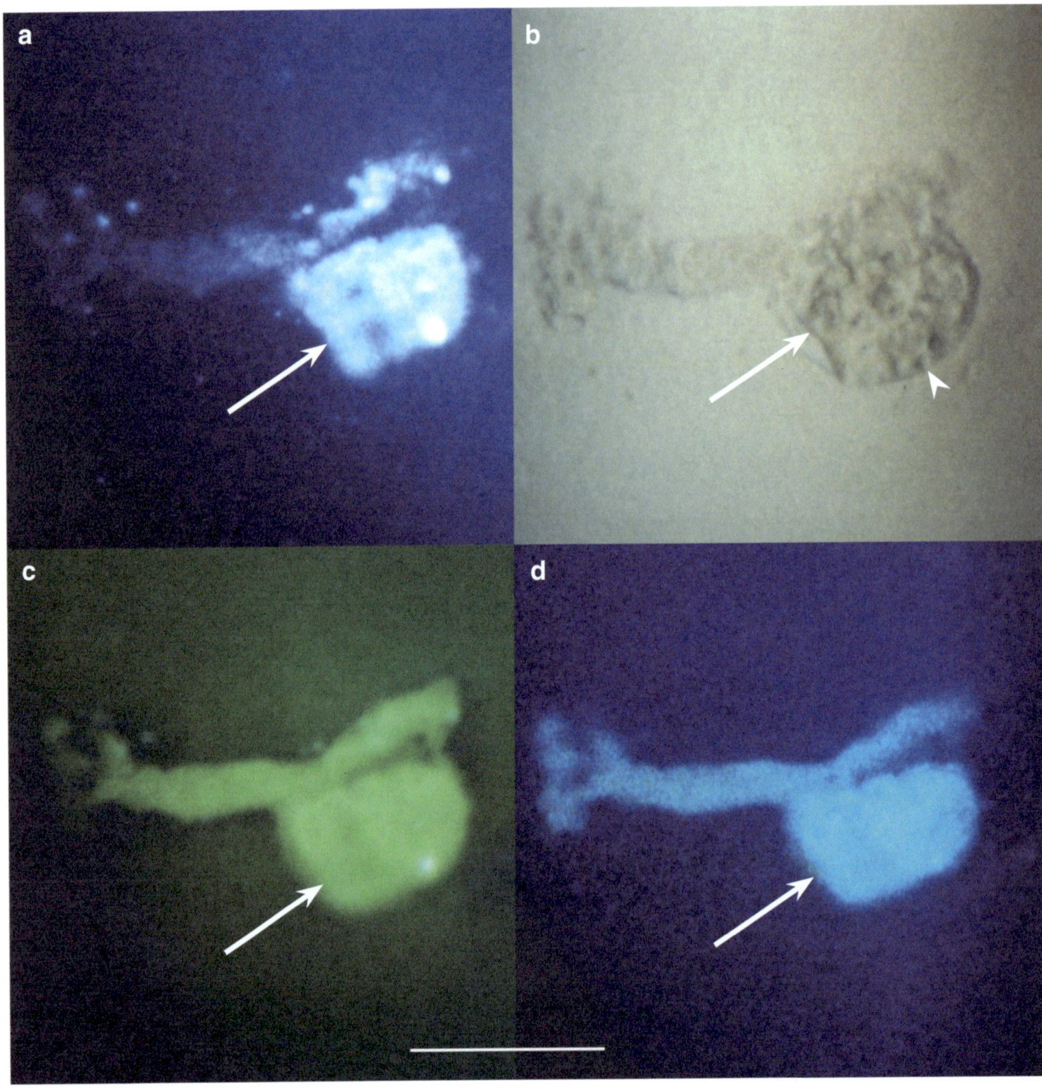

Fig. 1.37 (a–d) A lesion (*arrow*) with bizarre appearance. Its contents are (a) light-reflecting and (b) granular; the *arrowhead* indicates a rounded body. (c) The lesion stains green with fluorescein; the staining is sharply delineated and (c, d) there is no diffusion of the dye into the surroundings (the arrows are placed in corresponding locations; d, blue filter)

Complex Lesions (2/4)

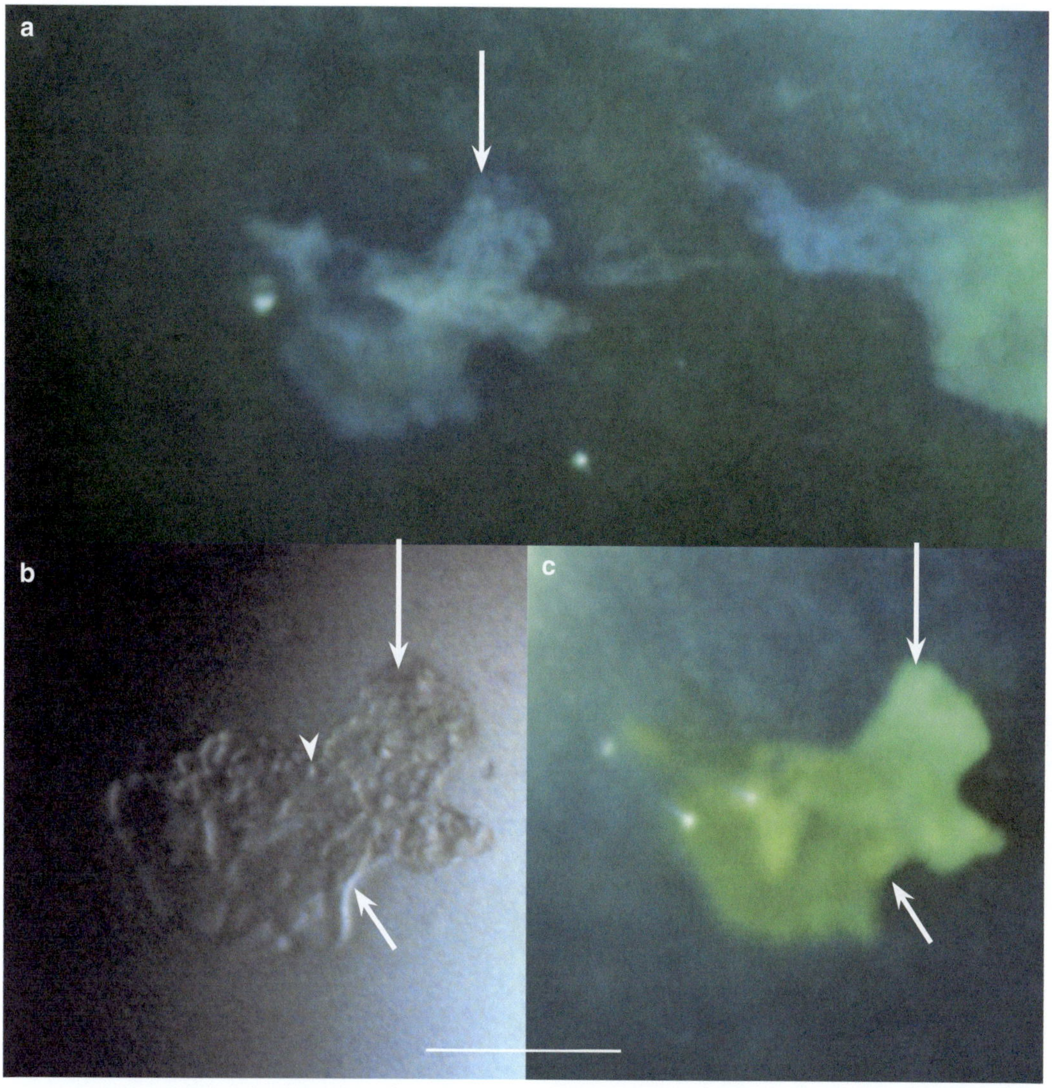

Fig. 1.38 (a) To the right of this light-reflecting not staining lesion (*arrow*) another one is visible which partly stains green with fluorescein. (b, c) The same lesion (*long arrow*) 24 h later. In (b) its granular contents are visible; the *arrowhead* indicates a rounded body. (c) shows various hues of the green fluorescein staining; the lesion is sharply delineated and there is no diffusion of the dye into the surroundings. The *short arrows* in (b, c) indicate a centripetally convex line implying a curved edge of an epithelial sheet (cf. Figs. 1.17, 1.18, 1.19 and 1.20) (the markers are placed in corresponding locations)

Complex Lesions (2/5)

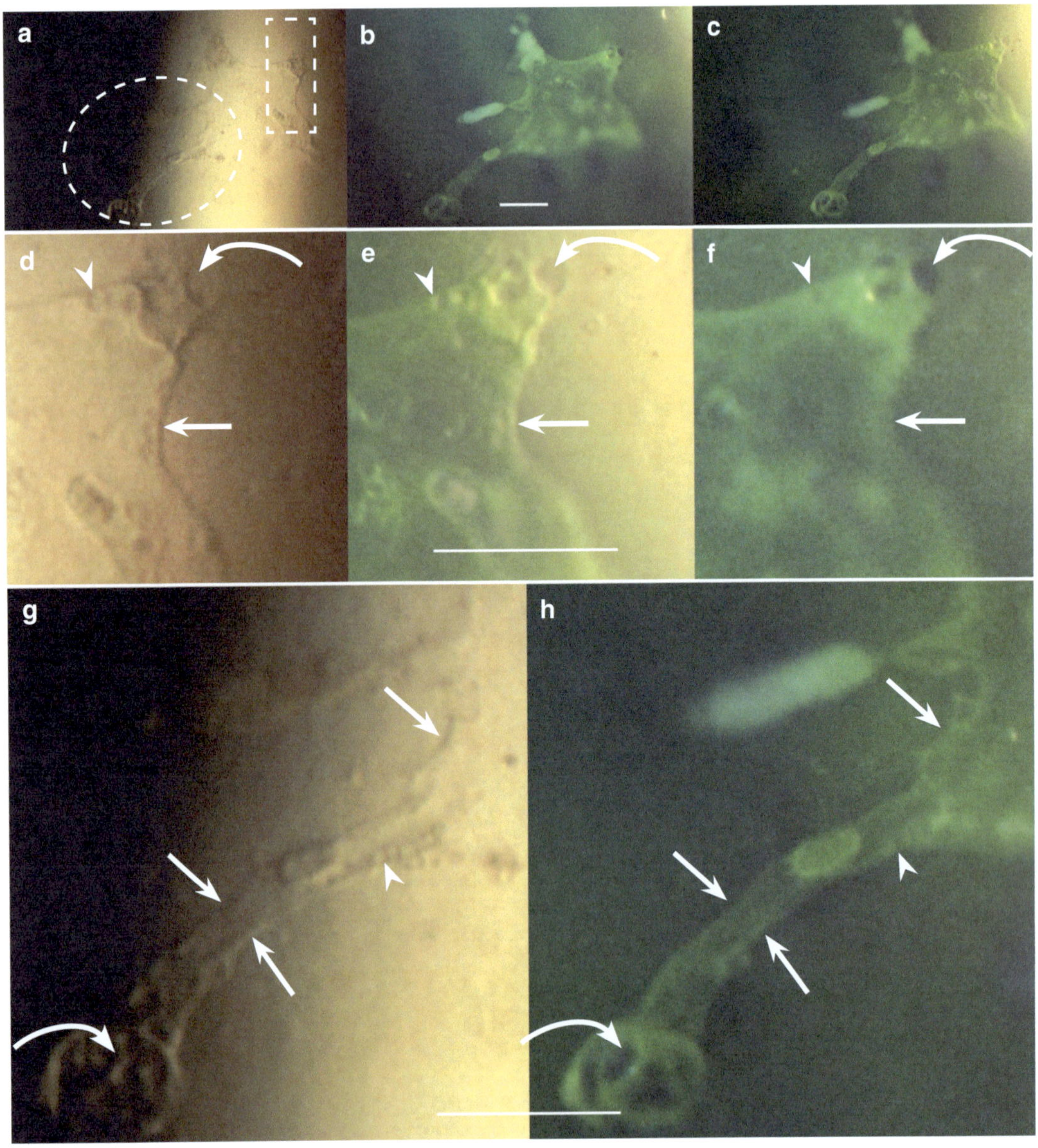

Fig. 1.39 (a–c) Survey of a lesion captured (a) before and (b, c) after the application of fluorescein. (d–f) The area indicated by *rectangular frame* in (a) shows a centripetally convex edge (*straight arrows*), rounded bodies (*arrowheads*) and small, smooth, rounded and elevated structures/islands (*bowed arrows*). (g, h) This area, indicated by *oval frame* in (a), shows smooth curved edges (*straight arrows*); in the lower part of the picture, they seem to be approaching each other. The *arrowheads* indicate rounded bodies and the *bowed arrows* small, smooth, rounded and elevated structures/islands within the lesion. In (f, h) the various hues of green fluorescein staining imply level differences (the markers are placed in corresponding locations in each set of photographs)

Complex Lesions and Their Surroundings (1)

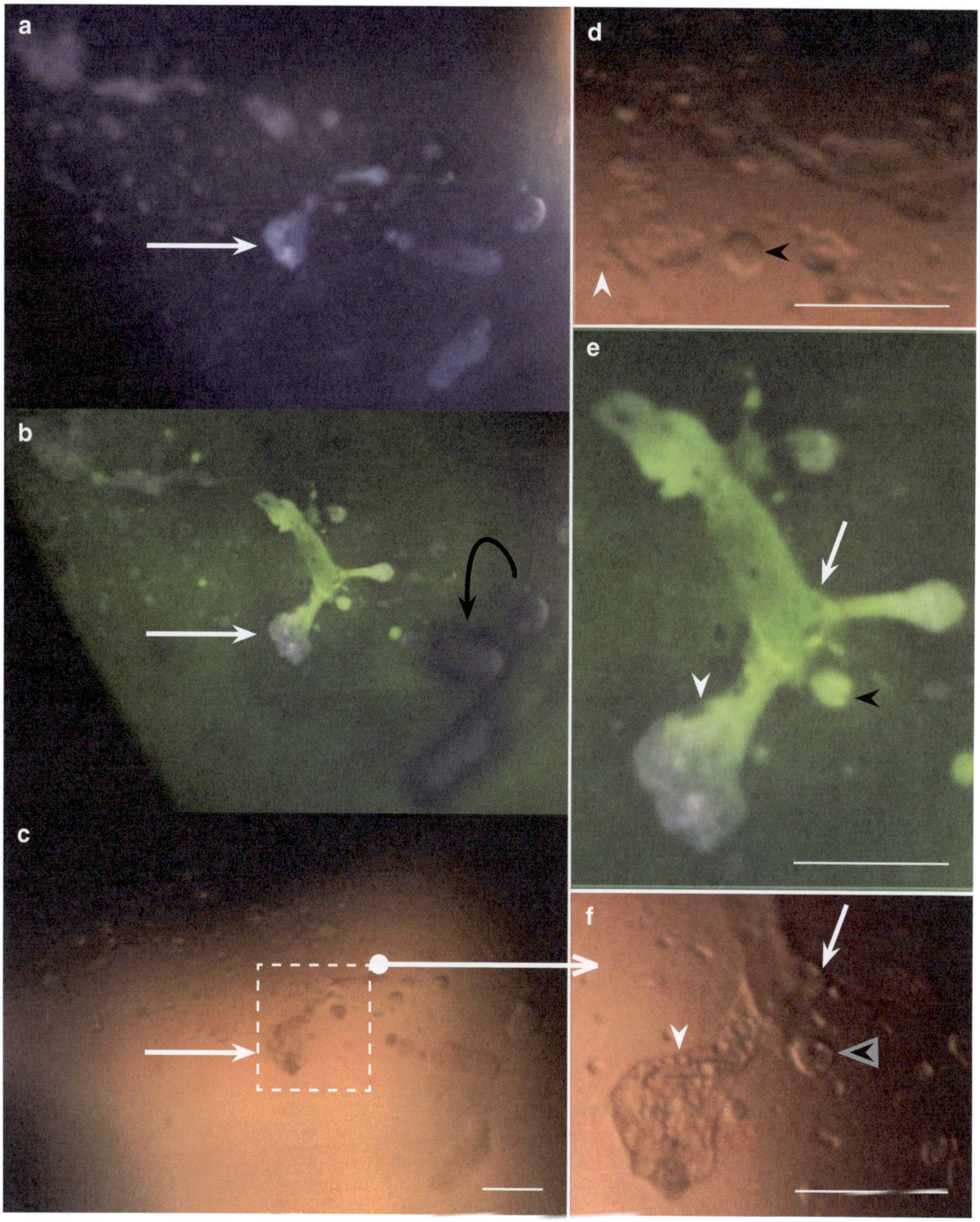

Fig. 1.40 (a–c) Survey of an area showing (a) many light-reflecting structures (*straight arrow*). (b) After the application of fluorescein appears a partly green stained lesion (*straight arrow*) and small green dots indicating cystic spaces; the light-reflecting structures to the right cause surface elevations (*bowed arrow*). (c) The same area in retroillumination. (The straight arrows are placed in corresponding locations.) (d) An area outside the lesion showing many cysts (*black arrowhead*) and rounded bodies (*white arrowhead*). (e) The lesion shows various green fluorescein hues. The *arrow* indicates a curved edge, the *black arrowhead* a cyst and the *white arrowhead* rounded bodies appearing dark, cf. (f) Left lower part of the lesion in retroillumination; the markers indicate the same structures as in (e). Compare Fig. 1.35a, b

Complex Lesions and Their Surroundings (2)

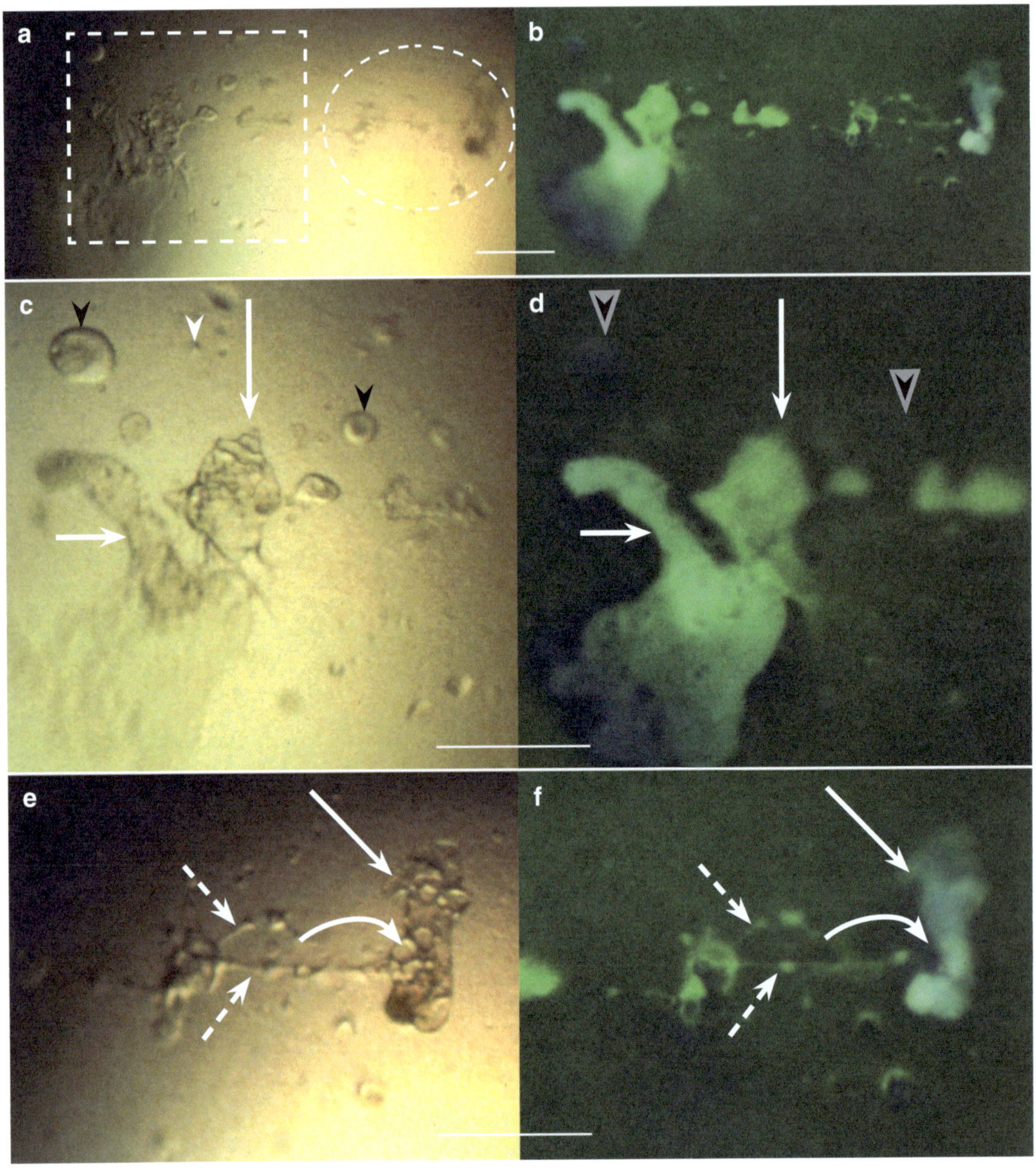

Fig. 1.41 (a, b) Survey of a deranged area (a) before and (b) after the application of fluorescein. (c, d) The area indicated by *rectangular frame* in (a) shows a larger lesion (*long arrow*) sharply delineated by curved edges (*short arrow*); cysts (*black arrowheads*); and rounded bodies (*white arrowhead*). (d) The green staining of the lesion (*long arrow*) shows various hues; its left lower part is not stained (cf. Fig. 1.26). The cysts (*black arrowheads*) do not stain. (e, f) The area indicated by *oval frame* in (a). In (e) are visible short epithelial lines (*dashed arrows*) with rounded expansions and a small lesion (*long arrow*) containing rounded bodies and small, smooth, rounded structures (*bowed arrow*). (f) With fluorescein, the lines (*dashed arrows*) and their expansions and some parts of the lesion stain green. Compare also Fig. 1.36 (the markers are placed in corresponding locations in each pair of photographs)

Complex Lesions and Their Surroundings (3)

Fig. 1.42 An area showing two larger lesions (*long arrows*) and their surroundings (**a**) before and (**b**) after the application of fluorescein. The *dashed arrows* indicate epithelial lines, the *white arrowheads* (in (**a**)) rounded bodies and the *black arrowheads* a cyst. In (**b**) are visible various hues of the green fluorescein staining (the arrows and the black arrowheads are placed in corresponding locations)

A Peculiar Lesion

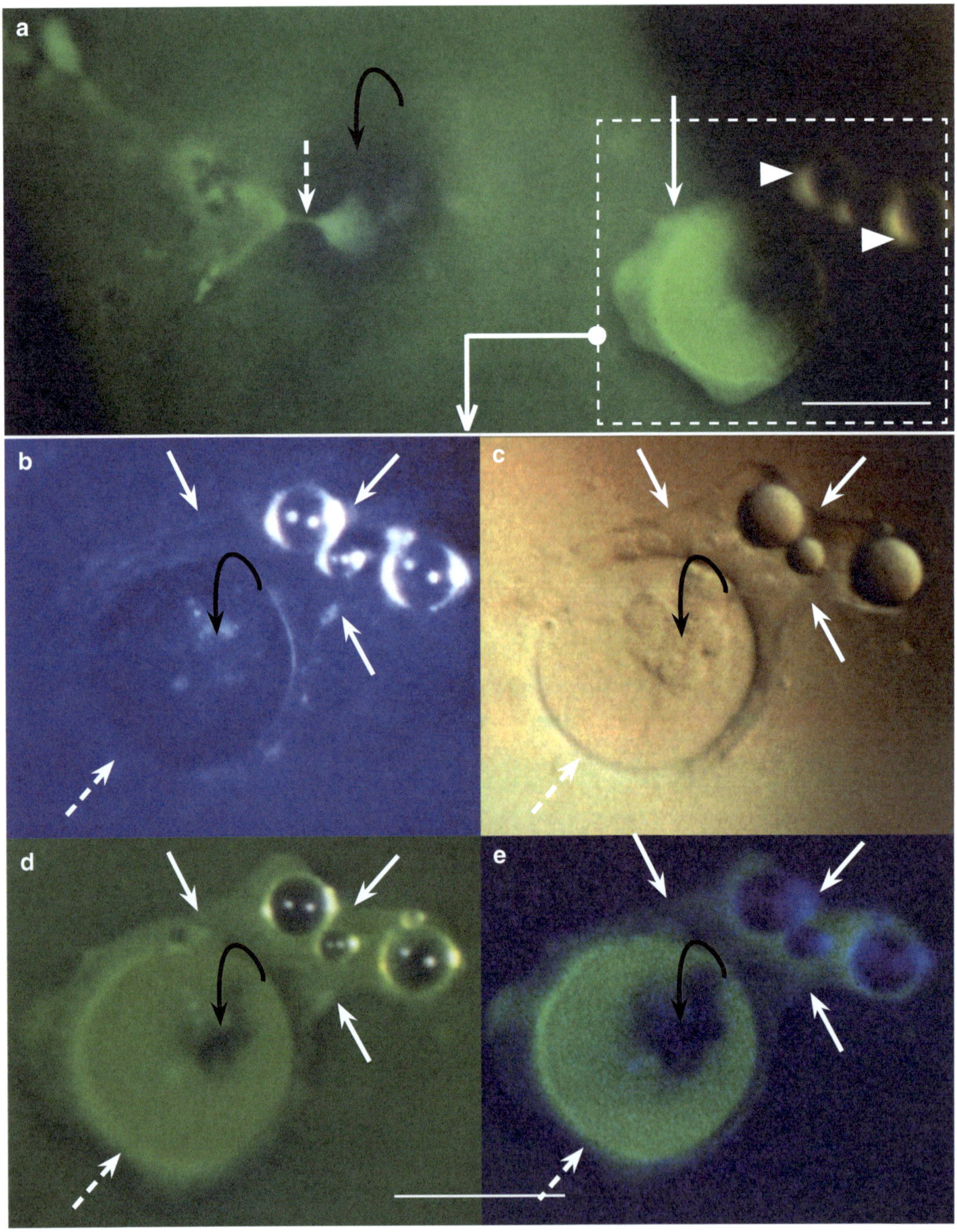

Fig. 1.43 (a) An area showing green stained epithelial lines (*dashed arrow*), surface elevations (dark, *bowed arrow*) and a peculiar lesion (*long arrow*); the *arrowheads* indicate entrapped air bubbles. (b–e) The lesion in *frame*, captured in (b) focal illumination, (c) retroillumination, (d) with fluorescein and (e) with fluorescein and blue filter. It has curved edges (*plain arrows*) and contains a circular area (*dashed arrow*) that shows a (b) partly light-reflecting (c) irregular structure (*bowed arrows*). (d, e) Fluorescein is pooling within the lesion; the structure (*bowed arrows*) is protruding (the markers in (b–e) are placed in corresponding locations)

A Surface Erosion

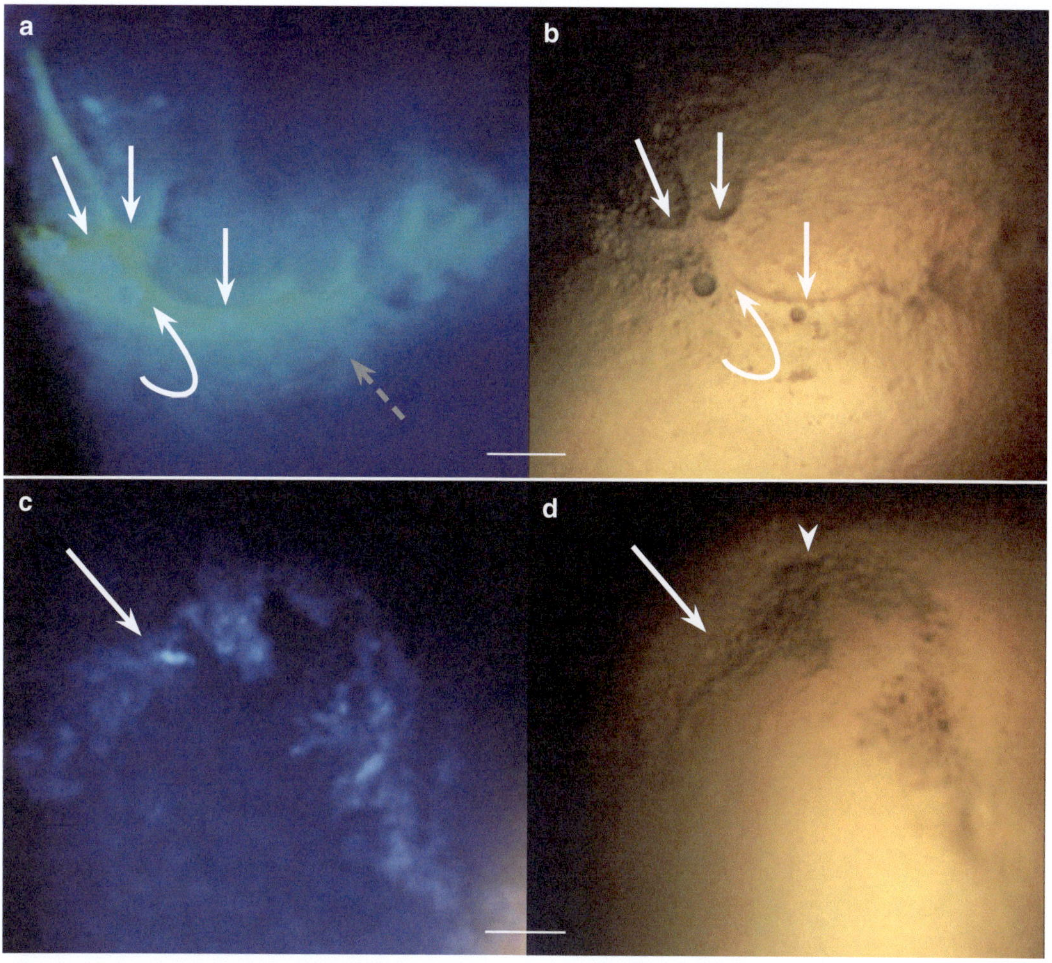

Fig. 1.44 (**a, b**) A spontaneous epithelial erosion in the lower part of the cornea about 24 h after symptom onset. Its upper edge shows several curved lines implying tips of epithelial sheets (*plain arrows*). The *bowed arrows* indicate a depressed area, in (**a**) visualised with fluorescein; the *dashed arrow* indicates fluorescein diffusion into the surroundings (**a**- blue filter). (**c, d**) The same location 24 h later. Visible is (**c**) a light-reflecting area (*arrow*) which (**d**) shows many heaped-up rounded bodies (*arrowhead*). (Post-traumatic recurrent erosion syndrome in the right eye for 5 years. The left cornea appeared normal. For additional examples of frank erosions, see Chap. 2, Cases 11 and 12)

Large, Well-Delineated Edematous Areas (1)

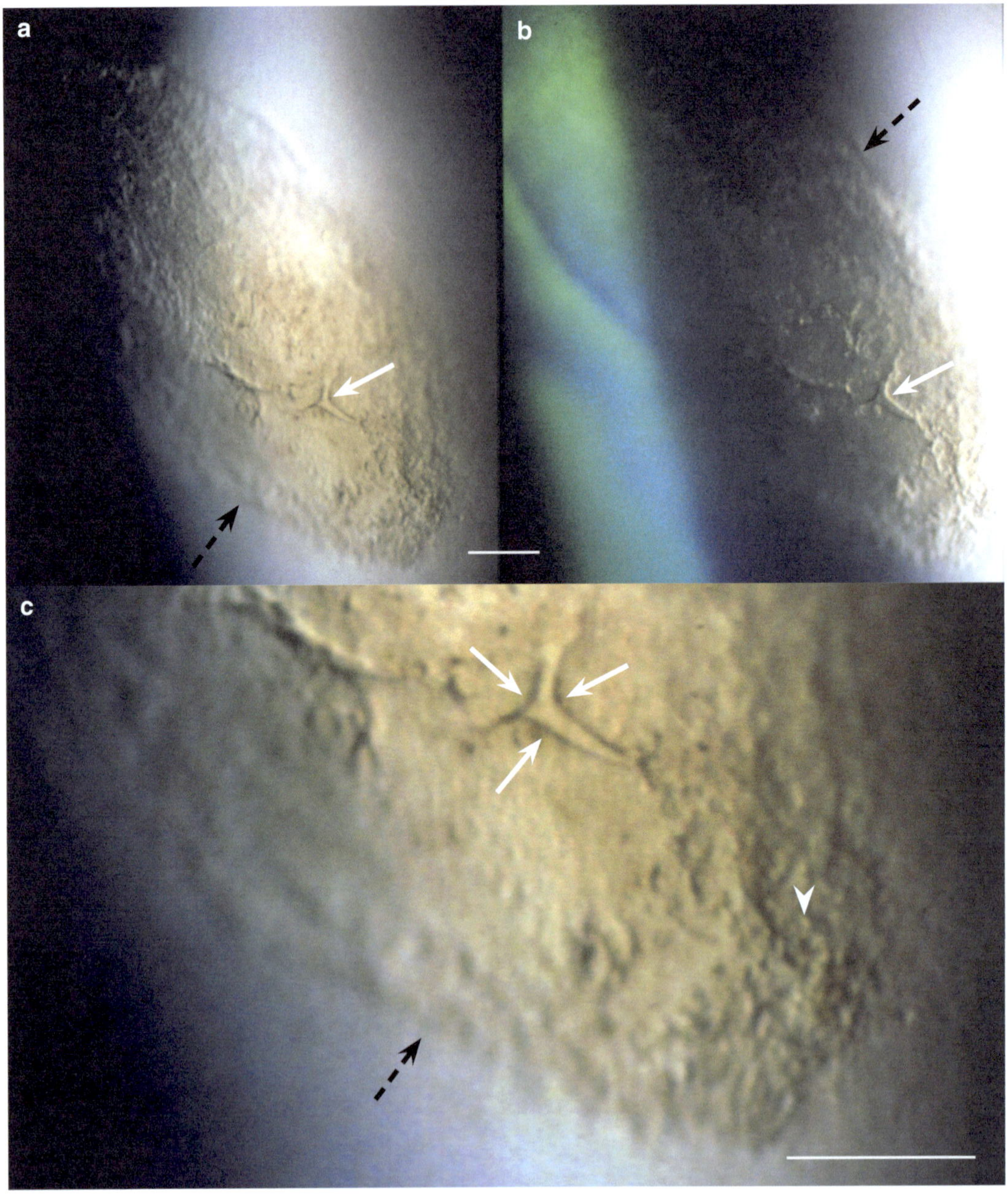

Fig. 1.45 (**a**, **b**) An oval, well-delineated (*black arrows*) edematous area with a small disruption showing centripetally convex edges (*white arrows*). Such areas usually stain green with fluorescein, cf. Fig. 1.1 (here, because of the illumination, captured only in its left upper part). (**c**) The lower part of the area at higher magnification. The *white arrows* indicate curved edges of epithelial sheets pointing toward the centre of the disruption and the *black arrow* the areas' edge (placed in the same locations as in (**a**)); the area shows many rounded bodies (*arrowhead*). Compare Fig. 1.46, opposite page

Large, Well-Delineated Edematous Areas (2)

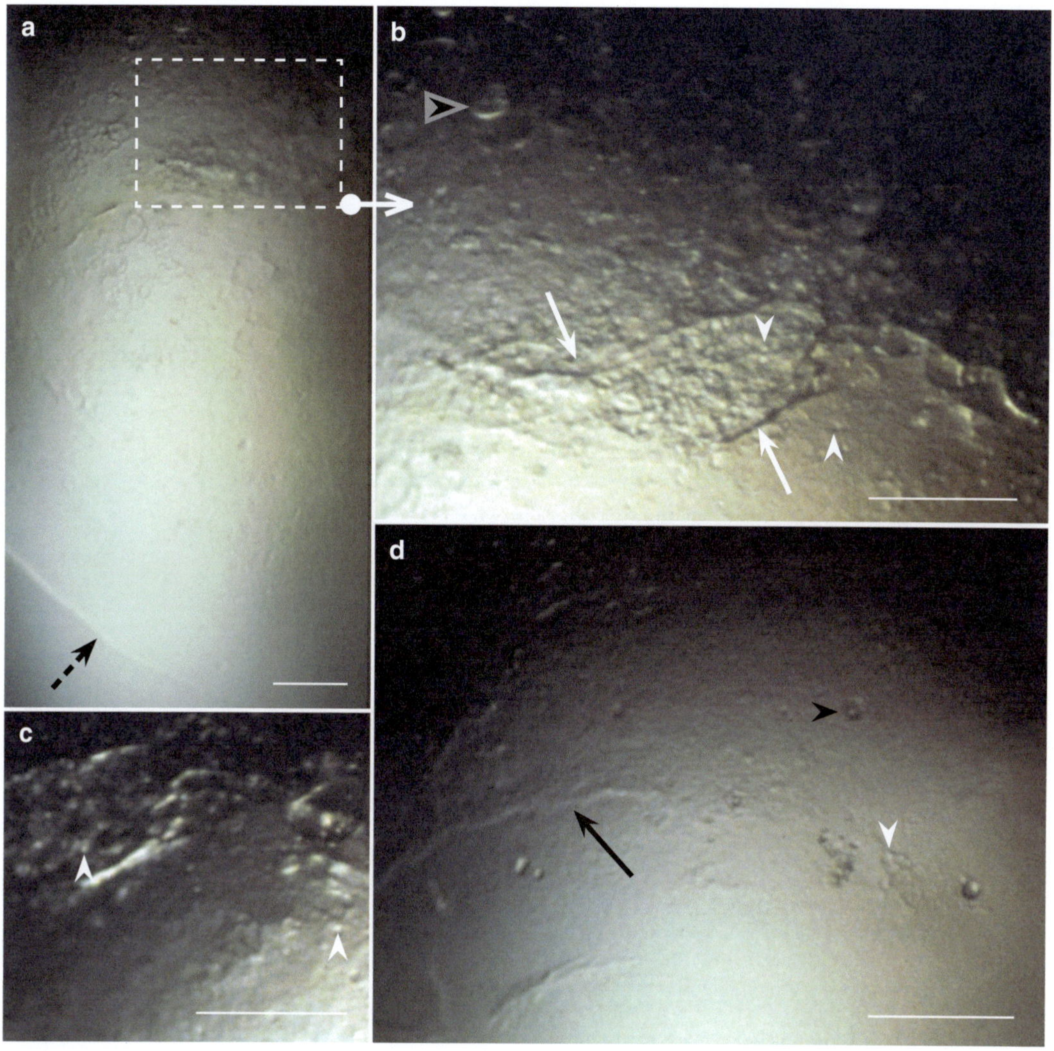

Fig. 1.46 (a) The *black arrow* indicates the left lower edge of an oval, well-delineated edematous area. (b) The part indicated by *frame* shows curved epithelial lines (*arrows*), many rounded bodies (*white arrowheads*) and a cyst (*black arrowhead*). Compare Fig. 2.33. (c, d) Rounded bodies (*white arrowheads*), a cyst (*black arrowhead*) and curved lines (*arrow*) probably representing epithelial wrinkles present within the edematous area

Addendum (1): Epithelial Cysts in a Corneal Scar (1)

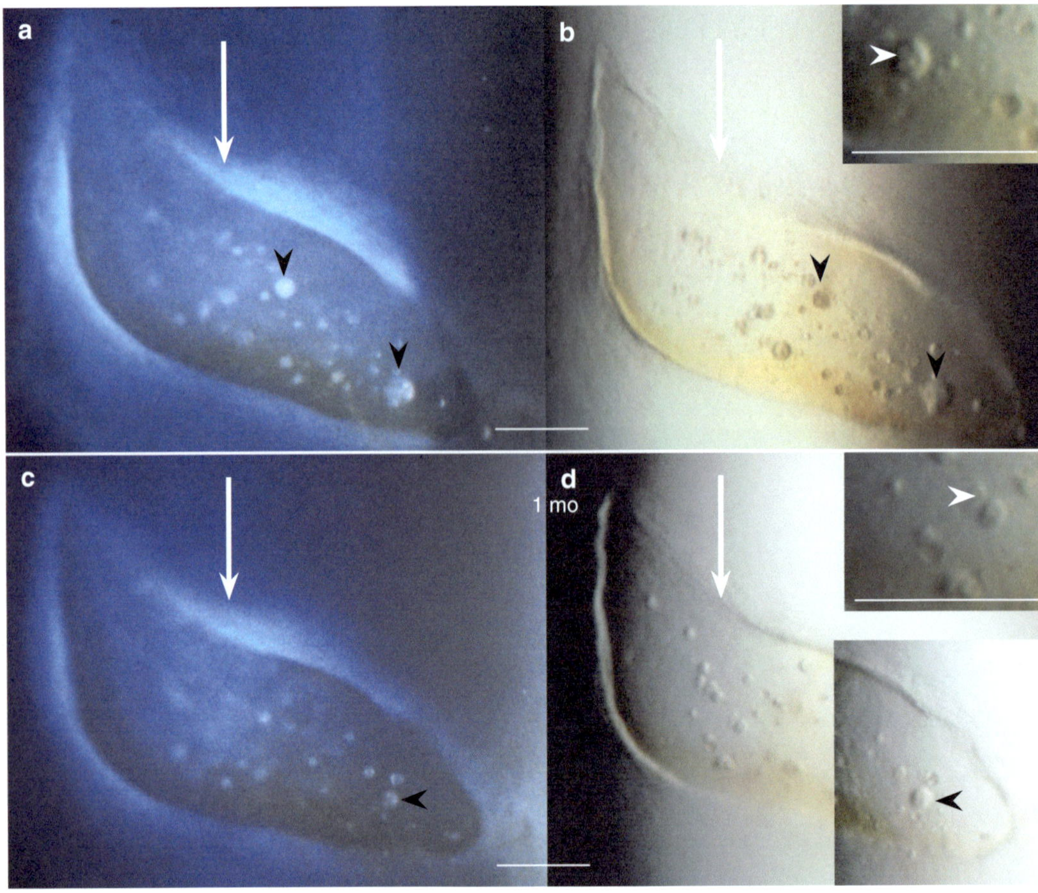

Fig. 1.47 Rounded epithelial cysts (*arrowheads*) situated in front of a subepithelial scar (*arrows*) after a metallic foreign body removed 17 years previously, (**a**, **b**) at presentation and (**c**, **d**) 1 month later. *Insets*: In some of the cysts rounded bodies (*arrowheads*) are visible (the arrowheads are placed in corresponding locations in each pair of photographs; (**d**) is a composed photograph) (Accidental finding. Three years before, the patient had a spontaneous erosion in the same area)

Addendum (2): Epithelial Cysts in a Corneal Scar (2) and in Keratoconus

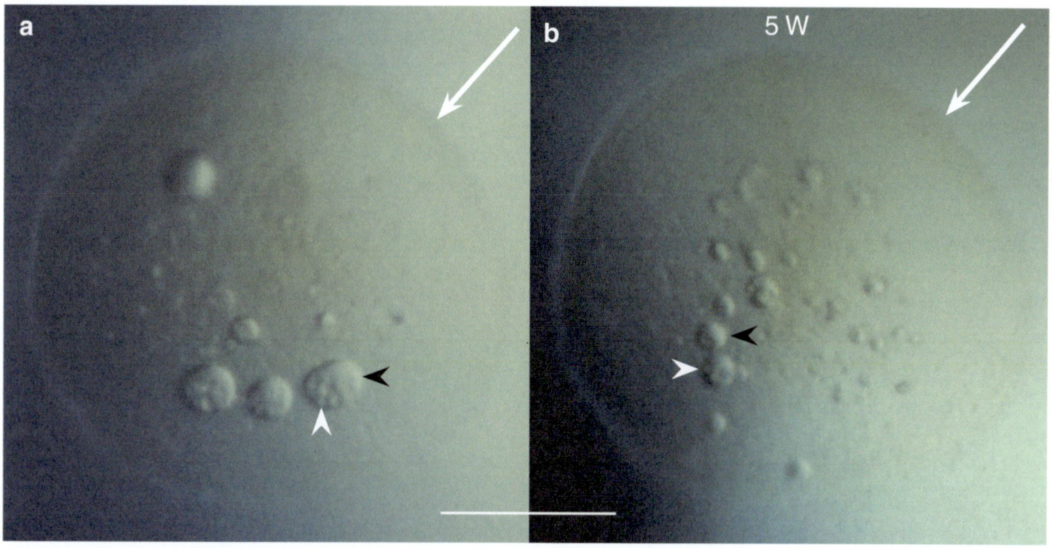

Fig. 1.48 Rounded epithelial cysts (*black arrowheads*), many containing rounded bodies (*white arrowheads*), situated in front of a subepithelial scar after a metallic foreign body removed about 30 years previously. (a) At presentation and (b) 5 weeks later. (Accidental finding in a symptom-free patient)

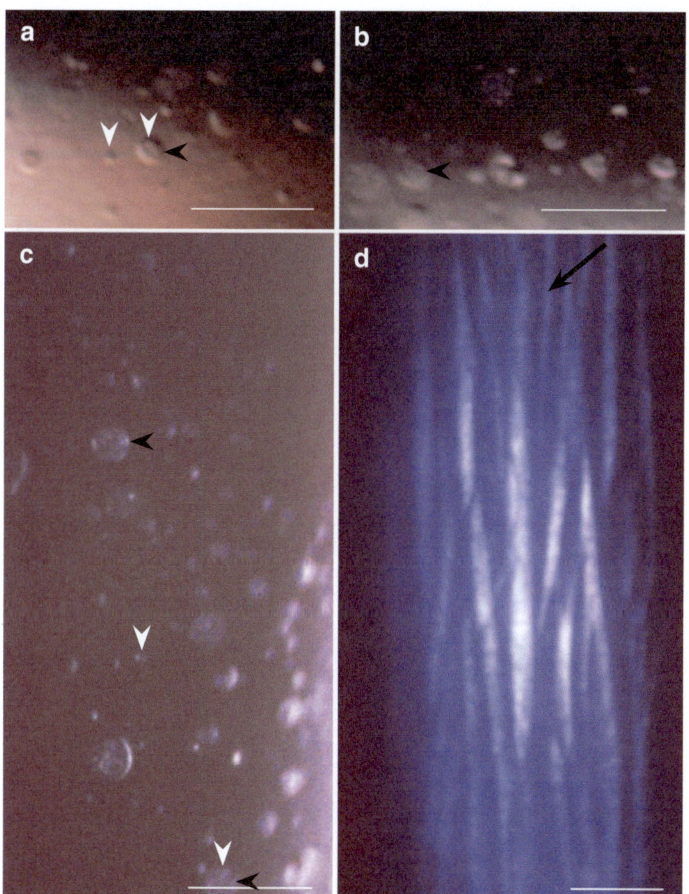

Fig. 1.49 (a–c) Rounded bodies (*white arrowheads*) outside and within some cysts (*black arrowheads*) in a cornea with keratoconus. (d) shows contemporaneously present vertical keratoconus (Vogt's) striae (*arrow*)

Case Reports

2

Twelve of the 13 cases included in this chapter partly illustrate the broad clinical spectrum of the recurrent erosion syndrome and partly its dynamic features as captured in the individual patient. In contrast, the last one (Case 13) demonstrates a cornea with pronounced epithelial basement membrane dystrophy in which repeated injuries did not result in recurrent erosions.

Eleven patients had typical recurrent erosion symptoms: three presented with frank erosions (Cases 1, 11 and 12), one with a well-delineated edematous epithelial area (Case 2) and seven with typical recurrent erosion findings (Cases 3, 4, 5, 6, 7, 8 and 9). In one patient (Case 10), the appearance of the corneal epithelium after a minimal trauma implied a predisposed cornea (see below).

Three patients (Cases 3, 4 and 5) had *KCS* but no corneal epithelial changes relatable to it. Rose bengal dye revealed some damaged superficial elements confined to deranged areas, which themselves were well discernible without staining or with addition of fluorescein. Case 3 additionally demonstrates the often encountered mixture of epithelial pathology and repair, Cases 4 and 6 the dynamics of fluorescein staining and Case 6 also a confusion with KCS symptoms.

Occasionally, recurrent erosions are confused with *herpes simplex* infections, possibly because both recur at irregular intervals and both show epithelial 'blisters'; their morphology, however, is entirely different. Case 7 stresses the importance of a careful slit-lamp observation in a patient having difficulty with verbal communication.

Sequential photographs revealed some *dynamic features* of the lesions. In one patient (Case 8) they showed a rapidly (within 4 h) enlarging part of a lesion. Whether this development was a natural event or whether the performed tonometry and the used anaesthetic (oxybuprocaine) were involved remains open.

The *dynamics of repair*, captured in two patients, illustrate two different situations. In Case 9 with a spontaneous episode of recurrent erosion, the resolution of a typical lesion visualises the origin of some of the commonly encountered phenomena. In Case 10, lacking history of recurrent erosions and reporting a minimal trauma, a well-delineated edematous area paralleling that often seen in the recurrent erosion syndrome implied a predisposed cornea, a suspicion supported by findings of cystic changes in the fellow cornea. The involved area contained many not light-reflecting rounded bodies, and its repair showed their successive elimination, a transient formation of fusiform expansions between approaching edges of healing epithelium before merging, and a division of the lesion in smaller parts the origin of which would be difficult to deduce if out of context.

The appearance of *frank epithelial erosions* after a mechanical injury is shown in two patients, both with known recurrent erosion syndrome. Twenty-four h (Case 11) and about 40 h (Case 12) after the injury, the corneae showed an extensive epithelial damage accompanied by stromal edema, endothelial deposits and flare in the anterior chamber. In one of them (Case 12), a later occurrence of a new, spontaneous episode demonstrates the difference in appearance in the same cornea of a frank erosion and epithelial disturbance typical of the recurrent erosion syndrome.

In Case 13 with pronounced *greyish membranes* (*maps*) but no recurrent erosion symptoms, sequential photographs performed during almost 7 years of follow-up captured dynamic features of maps and epithelial cysts.

H.M. Tabery, *Recurrent Erosion Syndrome and Epithelial Edema: In Vivo Morphology in the Human Cornea*, DOI 10.1007/978-3-319-06545-8_2, © Springer International Publishing Switzerland 2015

Case 1. Epithelial Erosion and Fingerprint Lines

Case Report

A 61-year-old man with spontaneous recurrent pain in the left eye on awakening for 2 months. At presentation, the left cornea showed disrupted surface with loose epithelial flaps and the right one many fingerprint lines. The erosion healed within 1 week.

The photographs of the left cornea were taken after a week's treatment with topical antibiotics and NaCl ointment 5 %.

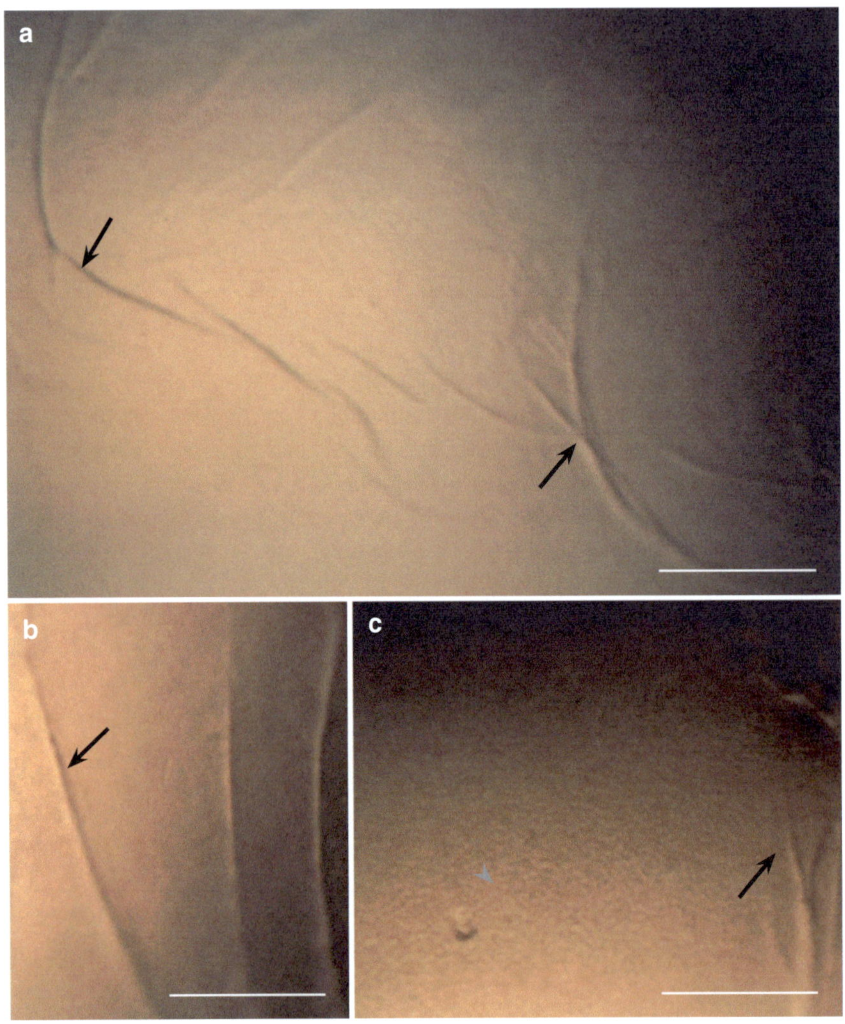

Fig. 2.1 (**a–c**) Fingerprint lines (*arrows*), in places crossing (**a**, *right arrow*). (**c**) additionally shows a fine granular pattern (*arrowhead*)

Case 2. Edematous Area, Fingerprint Lines and Blebs

Case Report

A 58-year-old man with spontaneous recurrent pain in the right eye on awakening for 2 weeks. At presentation, the paracentral inferior cornea showed an edematous, slightly elevated but not disrupted epithelial area, a few cysts, fingerprint lines and blebs. The left cornea showed discrete fingerprint lines. After a week's treatment with NaCl ointment 5 %, the edema disappeared and only a few small cysts were left. The photographs of the right cornea were taken at presentation.

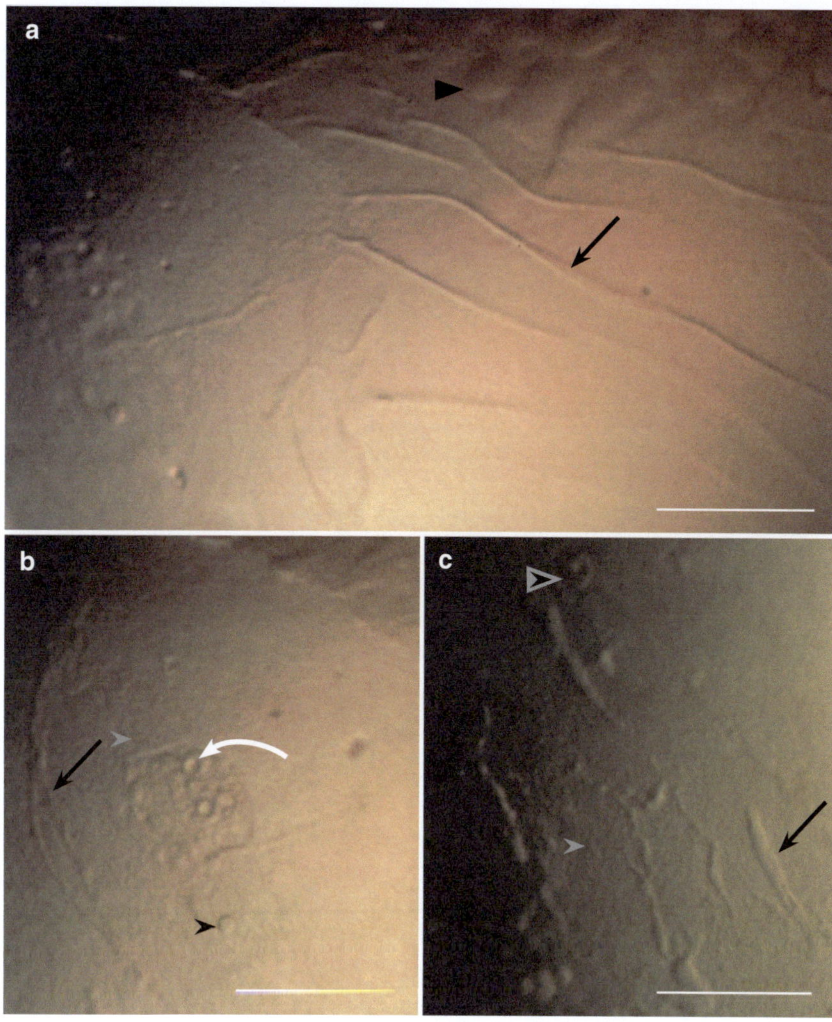

Fig. 2.2 Variously shaped fingerprint lines (*arrows*), blebs (**a**, *arrowhead*; cf. Fig. 1.12), cysts (**b**, **c**, *black arrowheads*), smooth rounded structures (**b**, *bowed arrow*) and a fine granular pattern (*grey arrowheads*)

Comment

The granular pattern in Figs. 2.1 and 2.2 probably represents intracellular edema (cf. Fig. 4.1).

Case 3. Recurrent Erosion and KCS (1)

Case Report

A 73-year-old woman with KCS and pain in the left eye on awakening for 2 months. At presentation, the left cornea showed a small lesion and many cysts and the right one fingerprint lines and a few cyst; there were no KCS-relatable changes. The symptoms subsided rapidly with lubricating eye drops and NaCl ointment 5 % at bedtime for a couple of weeks. Two years later, the same problem started in the right eye despite lubricating eye drops. The right cornea showed two lager lesions, cysts and fingerprint lines and the left one fingerprint lines and one small cyst. A lubricating ointment at bedtime was added. She was followed for a further 6 months. The symptoms in the right eye recurred but always subsided after an application of NaCl 5 % ointment at bedtime. When last seen, she was symptom-free.

The photographs of the symptomatic cornea were taken at respective presentation and of the right one also 1 month later.

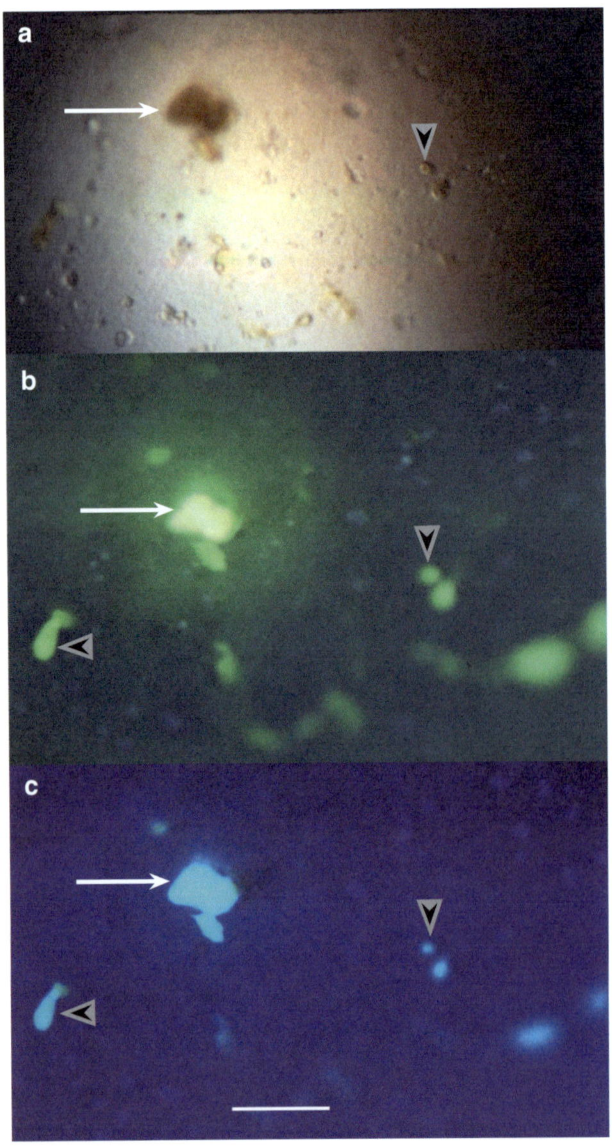

Fig. 2.3 Survey. (**a**) The deranged surface of the left cornea shows a small lesion (*arrow*) and many cysts (*arrowhead*). (**b**) The strong light reflection from the lesion (*arrow*) precludes the visibility in white light of the green fluorescein staining; many but not all cysts (*arrowheads*) stain green; fluorescein is still present in the tear film. (**c**) The same area visualised with fluorescein and blue filter. (The markers are placed in corresponding locations.) Some details of this area are shown in Fig. 2.4

Case 3. Recurrent Erosion and KCS (1) (cont.)

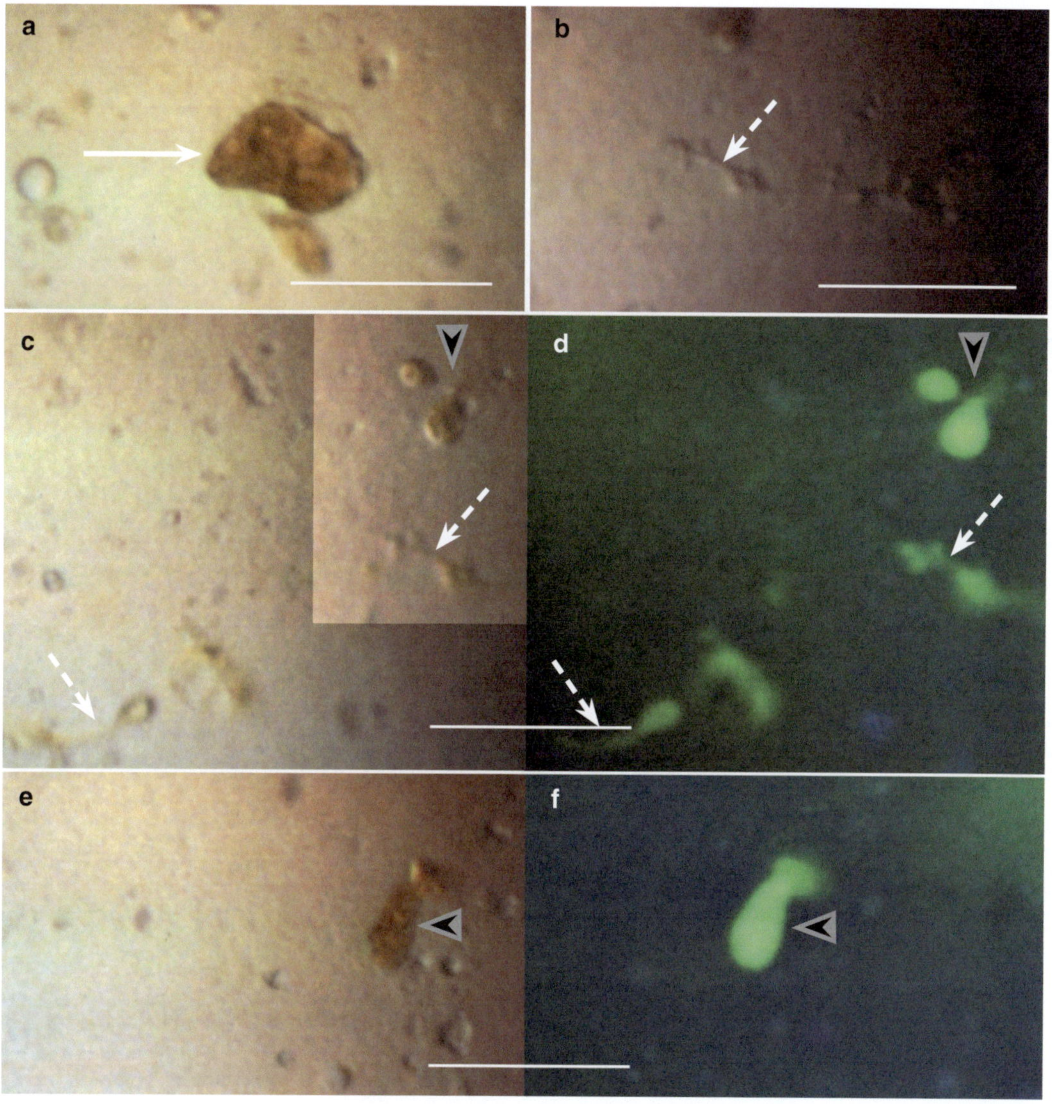

Fig. 2.4 (a) In retroillumination, fluorescein staining of the lesion (*arrow*) appears yellow-brown (cf. Fig. 2.5, overleaf). (b) A short epithelial line (*arrow*) between two small enlargements; in (d), the same line is stained green with fluorescein (*right arrow*). (c) Epithelial lines (*arrows*) and cystic spaces (*arrowheads*) staining (d) green with fluorescein. (e) Another cystic space (*arrowhead*) staining (f) green with fluorescein (the markers are placed in corresponding locations in each pair of photographs)

Comment

Epithelial lines and pear-shaped cystic spaces imply remnants of a recently present lesion (like that shown in Fig. 2.5, overleaf).

Case 3. Recurrent Erosion and KCS (1) (cont.)

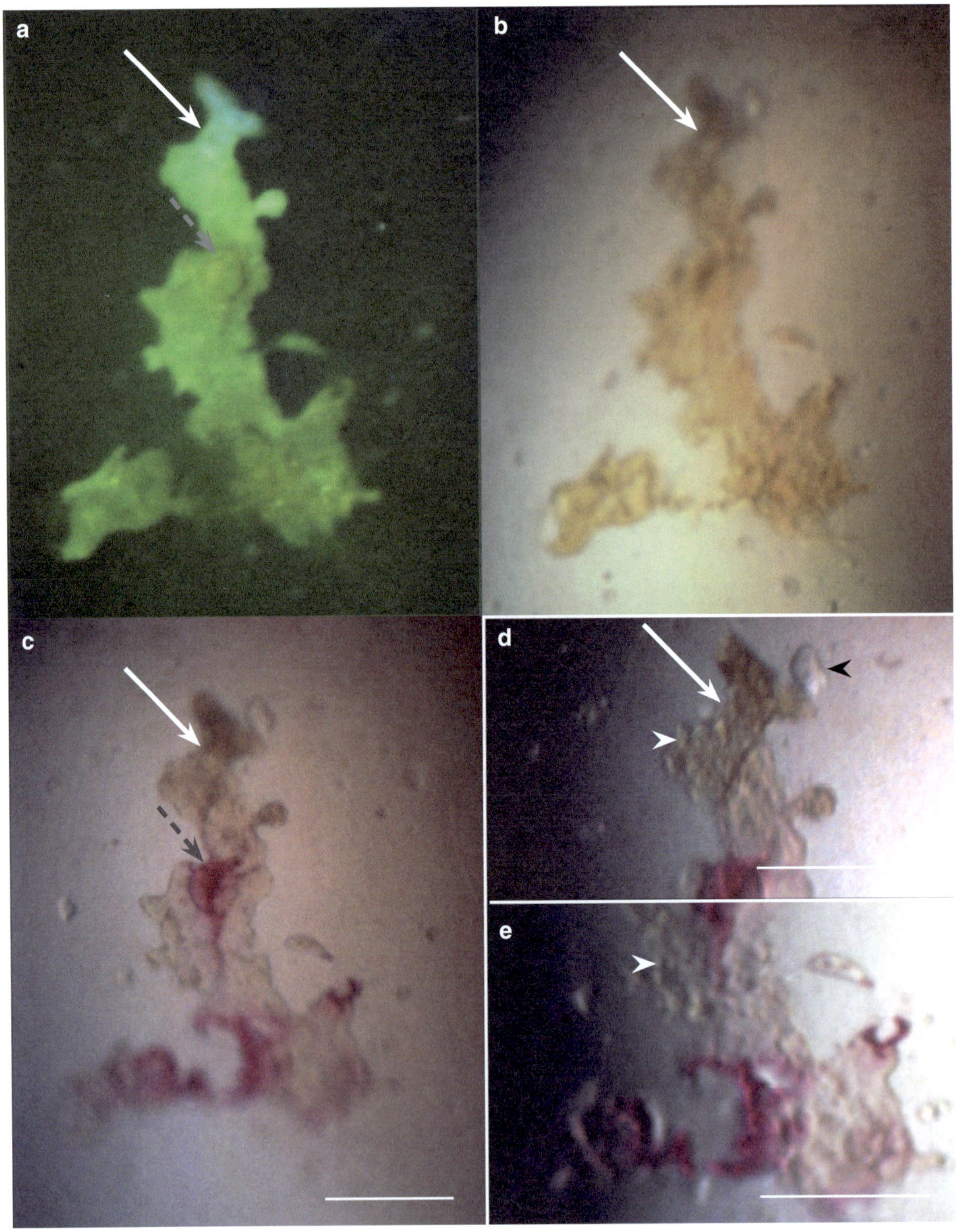

Fig. 2.5 (a–c) A large lesion (*long arrow*) present in the right cornea. (a) Fluorescein staining shows various shades of green. The *dashed arrow* indicates a circular area within the lesion (cf. Figs. 1.33 and 1.34). (b) In retroillumination, fluorescein staining appears yellow-brown. (c) The circular area (*dashed arrow*) and some other spots within the lesion stain red with rose bengal. (d–e) Both the upper (d) and the lower (e) part of the lesion (*arrow*) contain rounded bodies (*white arrowheads*). The cyst (d, *black arrowhead*) adjacent to the lesion does not stain with fluorescein (cf. a, b) (the arrows are placed in corresponding locations)

Case 3. Recurrent Erosion and KCS (1) (cont.)

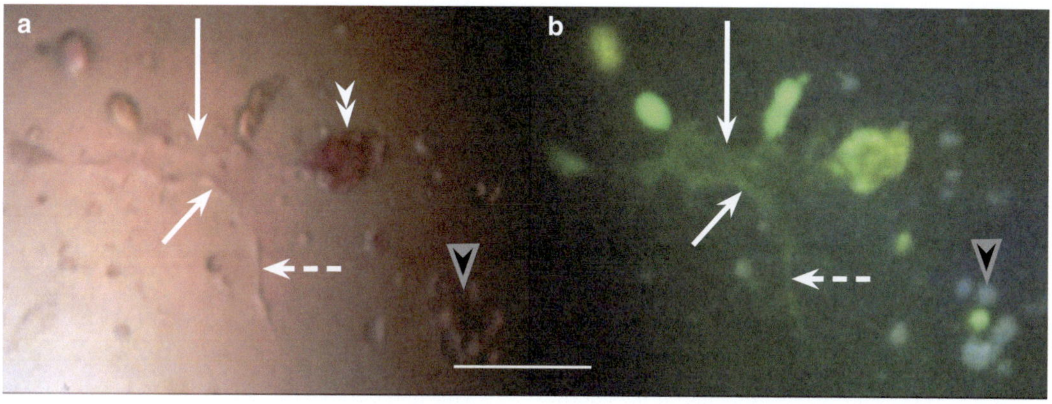

Fig. 2.6 (**a–b**) Another lesion (*long arrows*) captured in the right cornea at the same occasion as that in Fig. 2.5. The *short plain arrows* indicate a curved edge, the *dashed arrows* an epithelial line, the *double arrowhead* an area stained red with rose bengal and the *black arrowheads* a group of cysts. In (**b**) various hues of the green fluorescein staining of the lesion and green staining of some but not all cysts (*arrowhead*) are visible (cf. Fig. 2.23) (the arrows are placed in corresponding locations)

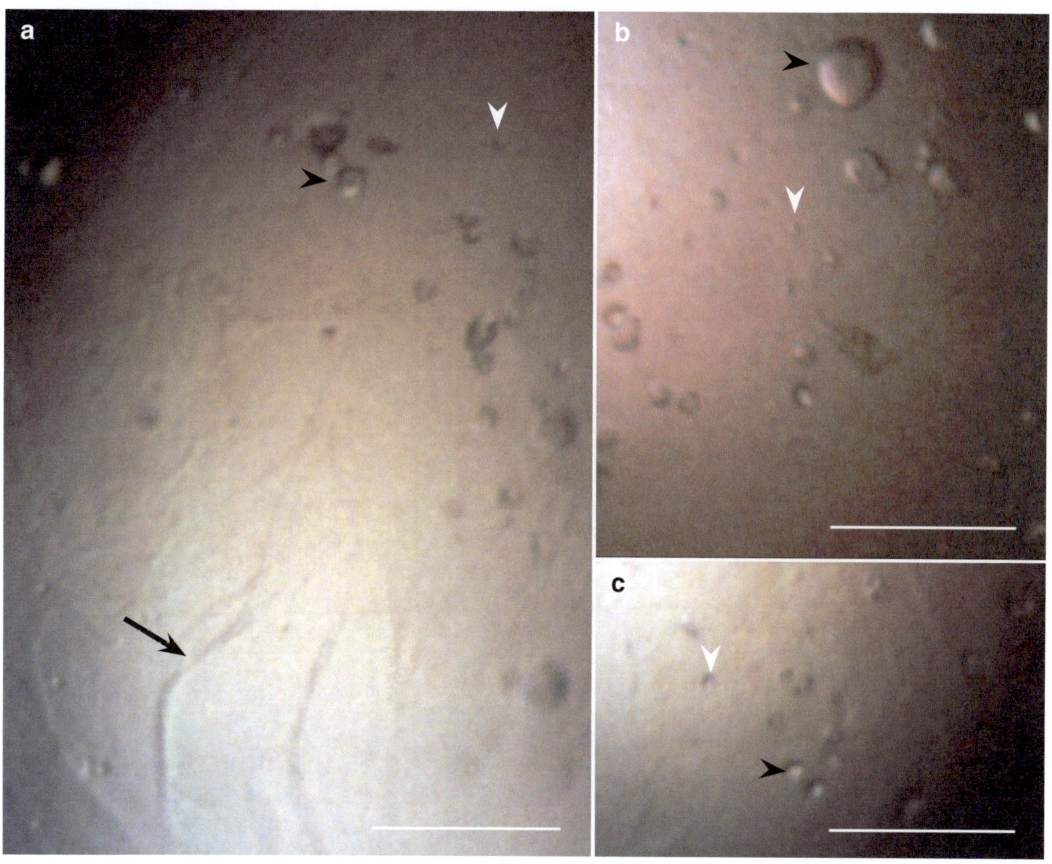

Fig. 2.7 (**a**) Fingerprint lines (*arrow*), (**a–c**) cysts (*black arrowheads*), and rounded bodies (*white arrowheads*) captured in the right cornea 1 month later

Case 4. Recurrent Erosion and KCS (2)

Case Report

A 53-year-old woman with recurrent pain in the right eye on awakening for several years, treated elsewhere with lubricating eye drops and ointment at bedtime, was referred for dry eye investigation. At presentation, the right cornea showed larger and smaller epithelial lesions and both corneae many small cysts. KCS tests supported the diagnosis, but the corneae showed no KCS-relatable changes. She was treated with lubricating eye drops and NaCl ointment 5 % at bedtime. One month later, she reported improvement although she had stopped the ointment after 2 weeks. After that, she was using lubricating eye drops and ointment sporadically. Three months after presentation, the cornea showed a larger lesion, but the symptoms were slight. During the follow-up, twice a year, the corneae either showed a few small epithelial cysts or appeared normal. When last seen, 6 years after the first visit, she had been symptom-free for 3 years.

The photographs of the right cornea were taken at presentation (day 1) and 1 month, 3 months and 16 months later.

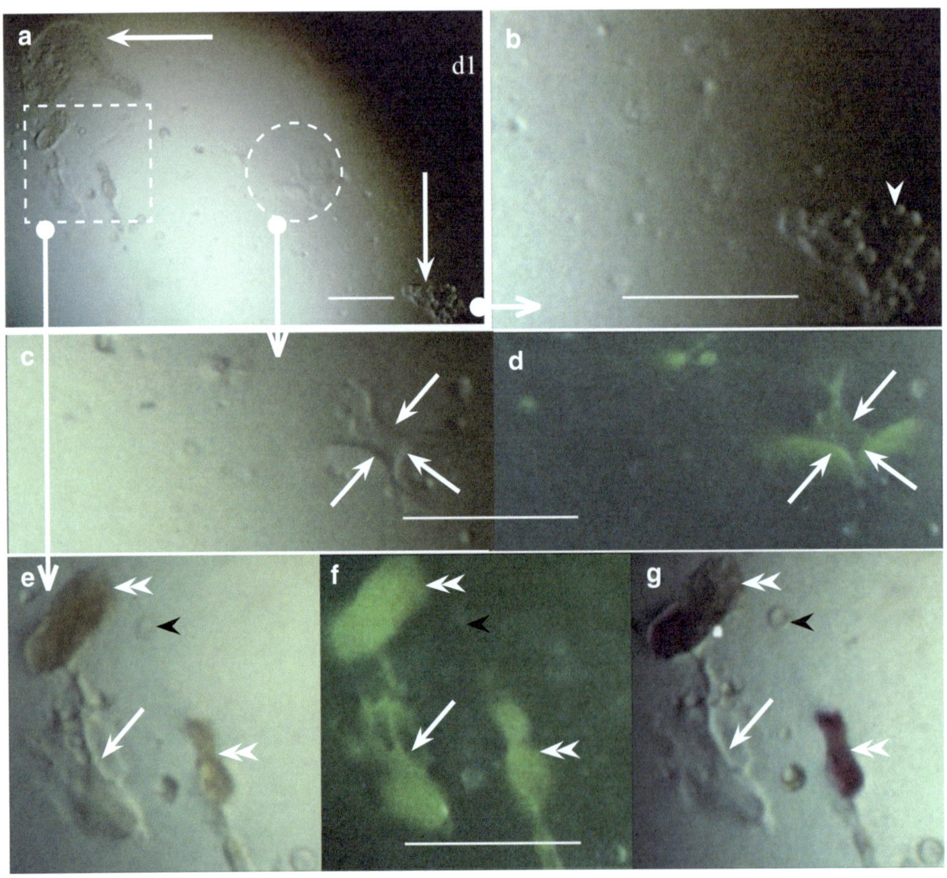

Fig. 2.8 (**a**) Survey of a deranged area containing a larger (*left arrow*) and a smaller (*right arrow*) lesion, which (**b**) contains many rounded bodies (*arrowhead*). (**c–d**) The area indicated by *round frame* in (**a**) shows approaching tips of epithelial sheets (*arrows*) and (**e–g**) that in *rectangular frame* two small adjacent lesions (*double arrowheads*) which stain (**e**) brownish and (**f**) green with fluorescein and (**g**) red with rose bengal; additionally are visible cysts (*black arrowheads*) and a line implying a small tip of an epithelial sheet (*arrow*). (The markers are placed in corresponding locations in each set of photographs.) For the dynamics of the green fluorescein staining, see Fig. 2.9

Case 4. Recurrent Erosion and KCS (2) (cont.)

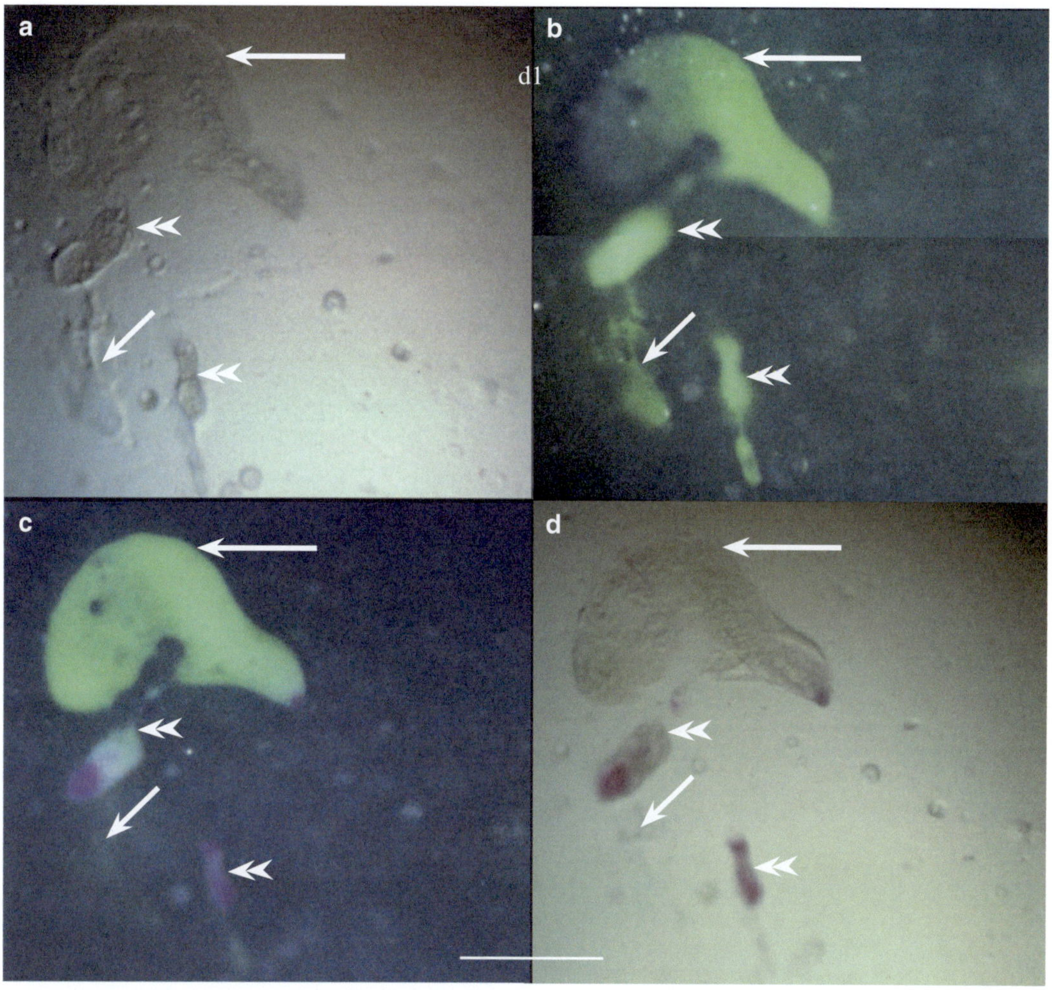

Fig. 2.9 (**a**) The large lesion (*long arrow*) stains (**b**) partly and after a while (**c**) completely green with fluorescein (cf. Fig. 2.11). (**b**, **c**) The green fluorescein staining indicated in (**b**) by *short arrow* has (**c**) disappeared. (**c**, **d**) Some surface elements stain red with rose bengal. The *double arrowheads* indicate the same small lesions as in Fig. 2.8e–g. The photographs are shown in the order as taken. (**b**) is a composed photograph

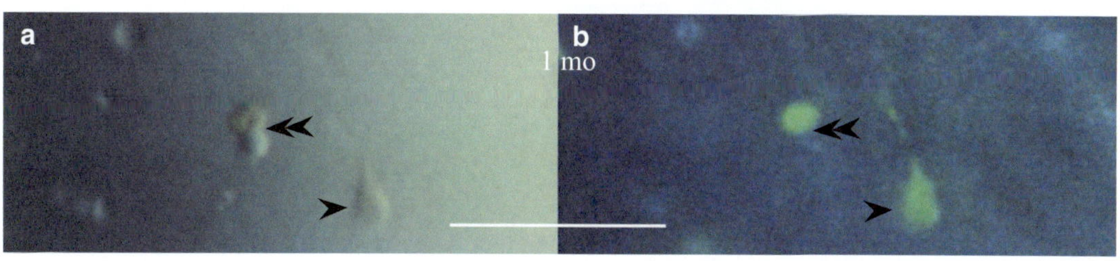

Fig. 2.10 (**a**) Small cysts that appear confluent (*double arrowhead*) but only the upper one stains (**b**) green with fluorescein. The *arrowhead* indicates a pear-shaped cyst (the markers are placed in corresponding locations)

Case 4. Recurrent Erosion and KCS (2) (cont.)

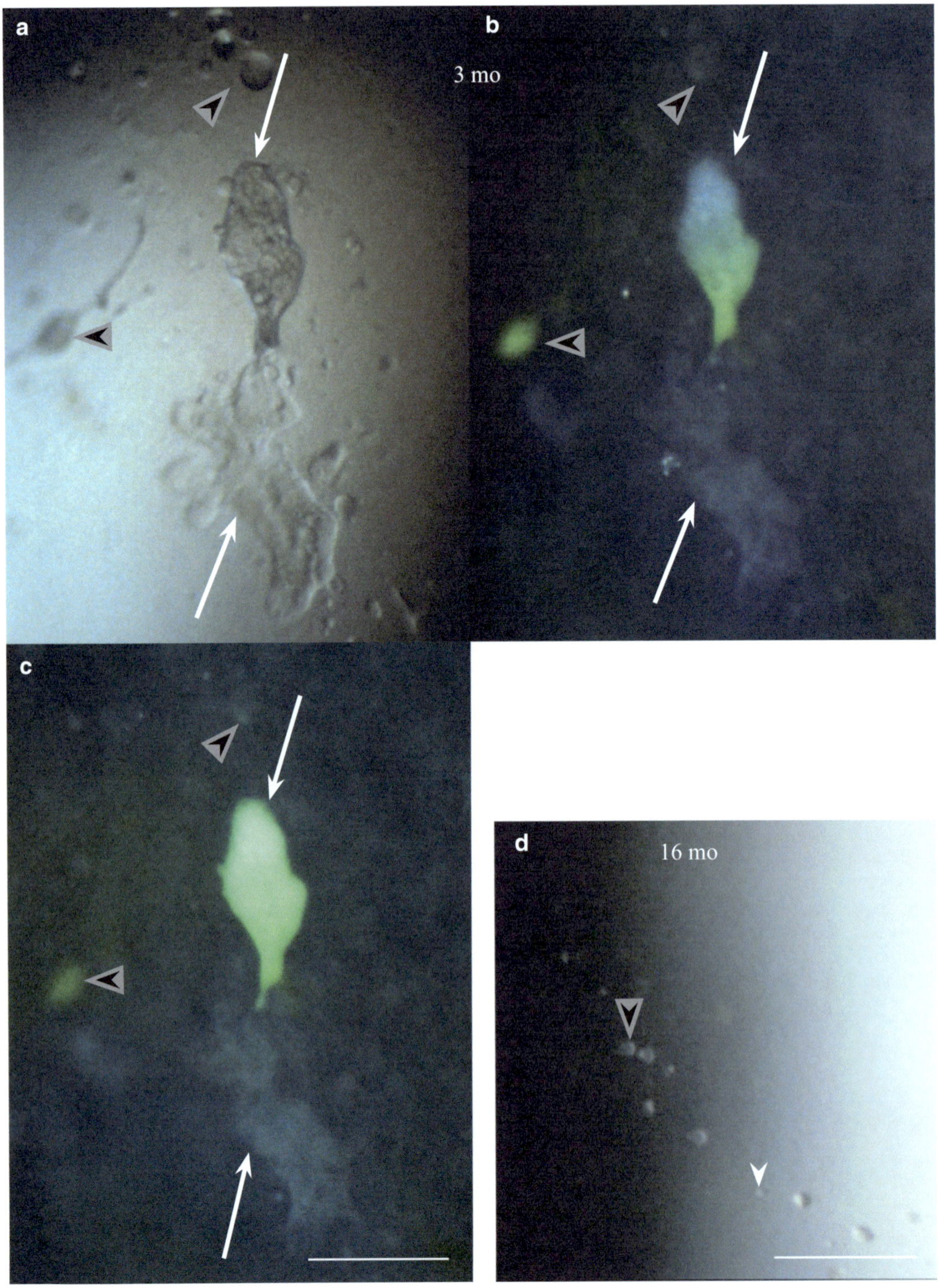

Fig. 2.11 (a–c) Two lesions (*arrows*) in apposition. The upper lesion stains (b) partly and (c) entirely green with fluorescein; the lower one does not stain (cf. Figs. 2.19, 2.21 and 3.2). The surroundings show cysts (*arrowheads*) of which some (b, c) stain green. (The photographs are shown in the order as taken. The markers are placed in corresponding locations.) (d) At this occasion, the cornea showed only a few small cysts (*black arrowhead*) and rounded bodies (*white arrowhead*)

Case 5. Recurrent Erosion and KCS (3)

Case Report

A 57-year-old woman under investigation for Sjögren's syndrome with grittiness in both eyes on awakening for many years but no symptoms during daytime. Lubricating eye drops had no effect. At presentation, KCS tests supported the diagnosis. The conjunctivae showed pronounced rose bengal staining but none was present on the corneae. Both corneae showed many epithelial cysts, fingerprint lines and greyish membranes ('maps') and the right cornea additionally an epithelial lesion. A lubricating ointment at bedtime was added; the symptoms persisted but were relieved by NaCl ointment 5 % at bedtime. During the observation period of almost 2 years, KCS tests showed the same results and both corneae showed the same changes as before, but no new epithelial lesions were observed.

The photographs of the right eye were taken at presentation (day 1) and 15 months later.

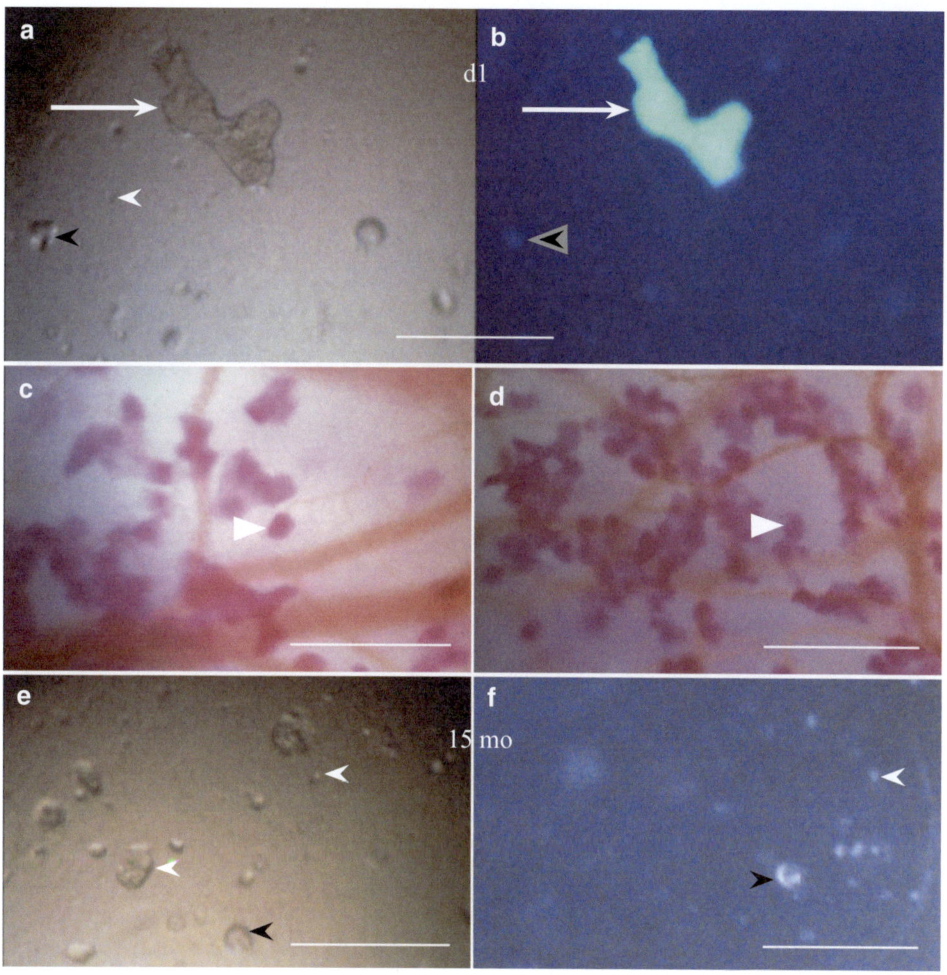

Fig. 2.12 (**a, b**) A lesion (*arrows*), cysts (*black arrowheads*) and rounded bodies (*white arrowhead*); in (**b**), the visibility of fluorescein staining is hampered by a strong light reflection from the lesion; the cyst (*arrowhead*) does not stain ((**b**) blue filter). (**c, d**) Rose bengal staining of diseased conjunctival cells (*arrowheads*). (**e, f**) Cysts (*black arrowheads*) and rounded bodies (*white arrowheads*) in (**e**) retroillumination and (**f**) focal illumination (not the same areas)

Case 6. Recurrent Erosions or KCS Symptoms?

Case Report

A 62-year-old woman with hypothyreosis after thyreoiditis, psoriasis, joint pain and foreign body sensation or pain in either eye on awakening for 6 years was treated for (presumed) dry eye with lubricating eye drops and topical cortisone. At presentation, KCS test values were well within normal range. The right cornea showed a larger epithelial lesion and the left one many small cysts. She was treated with lubricating eye drops and NaCl ointment 5 % at bedtime. Within 11 days, her symptoms had disappeared, but both corneae showed small cysts; 6 weeks after presentation, the right cornea additionally showed a small epithelial lesion. Five months after presentation both corneae showed small cysts.

The photographs of the right cornea were taken at presentation (day 1) and 11 days, 6 weeks, 10 weeks and 5 months later.

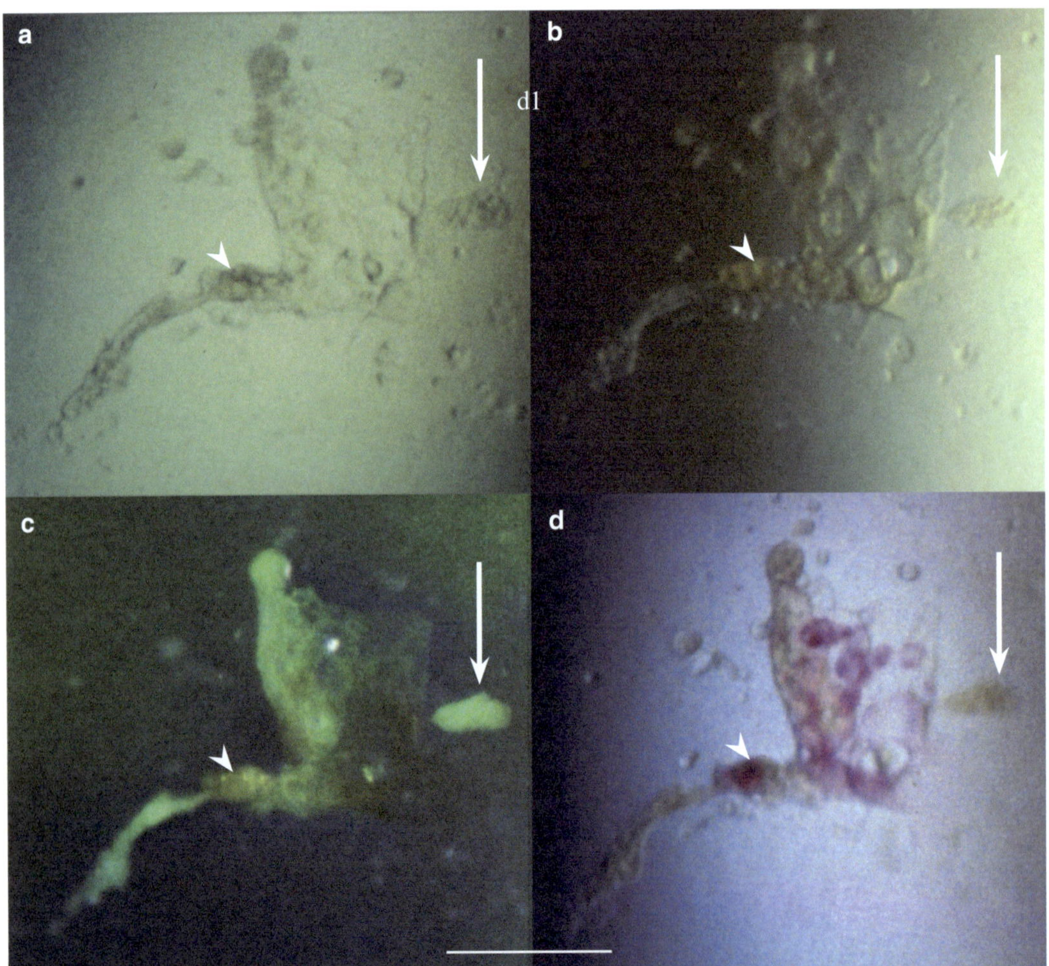

Fig. 2.13 A large lesion with bizarre shape and chaotic appearance and a small one in apposition (*arrow*). (**a, b**) The large lesion seems to consist of variously large and variously shaped compartments; the *arrowheads* indicate rounded bodies. (**c**) Early after the application of fluorescein, the large lesion shows various shades of green; its *upper right part* appears mottled, and in places such as at the *arrowhead* there is a brownish component. (**c**) Some parts of the lesion stain red with rose bengal; the small lesion does not. (The markers are placed in corresponding locations. See also Fig. 1.36e–d). For dynamics of the green fluorescein staining, see Figs. 2.14 and 2.15

Case 6. Recurrent Erosions or KCS Symptoms? (cont.)

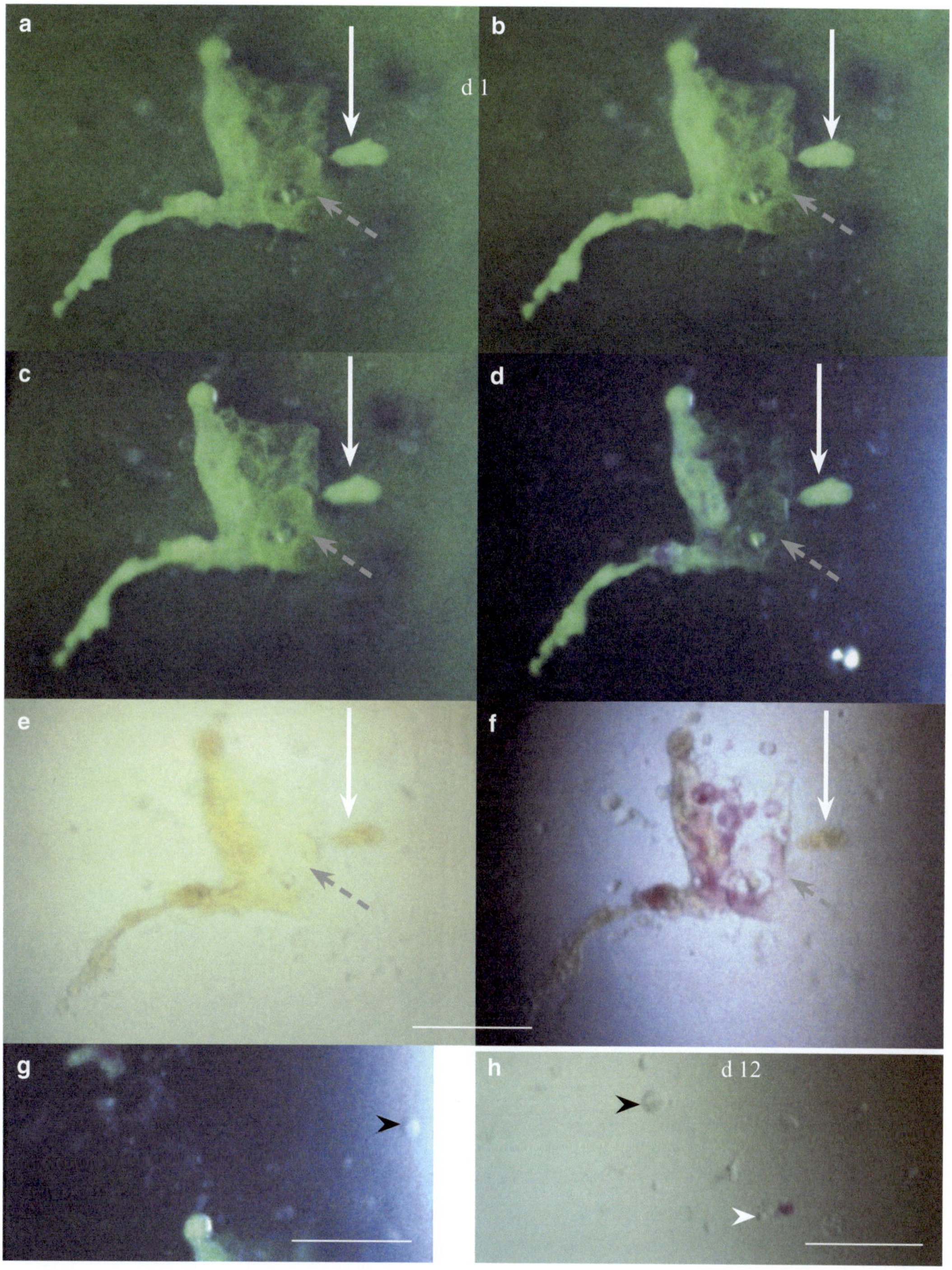

Fig. 2.14 (a–c) Brilliantly green staining of the small lesion (*white arrow*) and of the left part of the large one; its right upper part appears mottled and the lower one contains an oval area (*dashed arrow*; see also Fig. 2.25, d2). (d) After rose bengal staining, fluorescein has disappeared from the tear film but not from the lesions. (The photographs are shown in the order as taken.) (e) The green stained parts appear yellow in retroillumination. (f) Some parts of the lesion stain red with rose bengal. (g) A few cysts (*arrowhead*) and light-reflecting (greyish) structures adjacent to the upper part of the large lesion. (h) Only a few small cysts (*black arrowhead*) and rounded bodies (*white arrowhead*) were present at this occasion

Case 6. Recurrent Erosions or KCS Symptoms? (cont.)

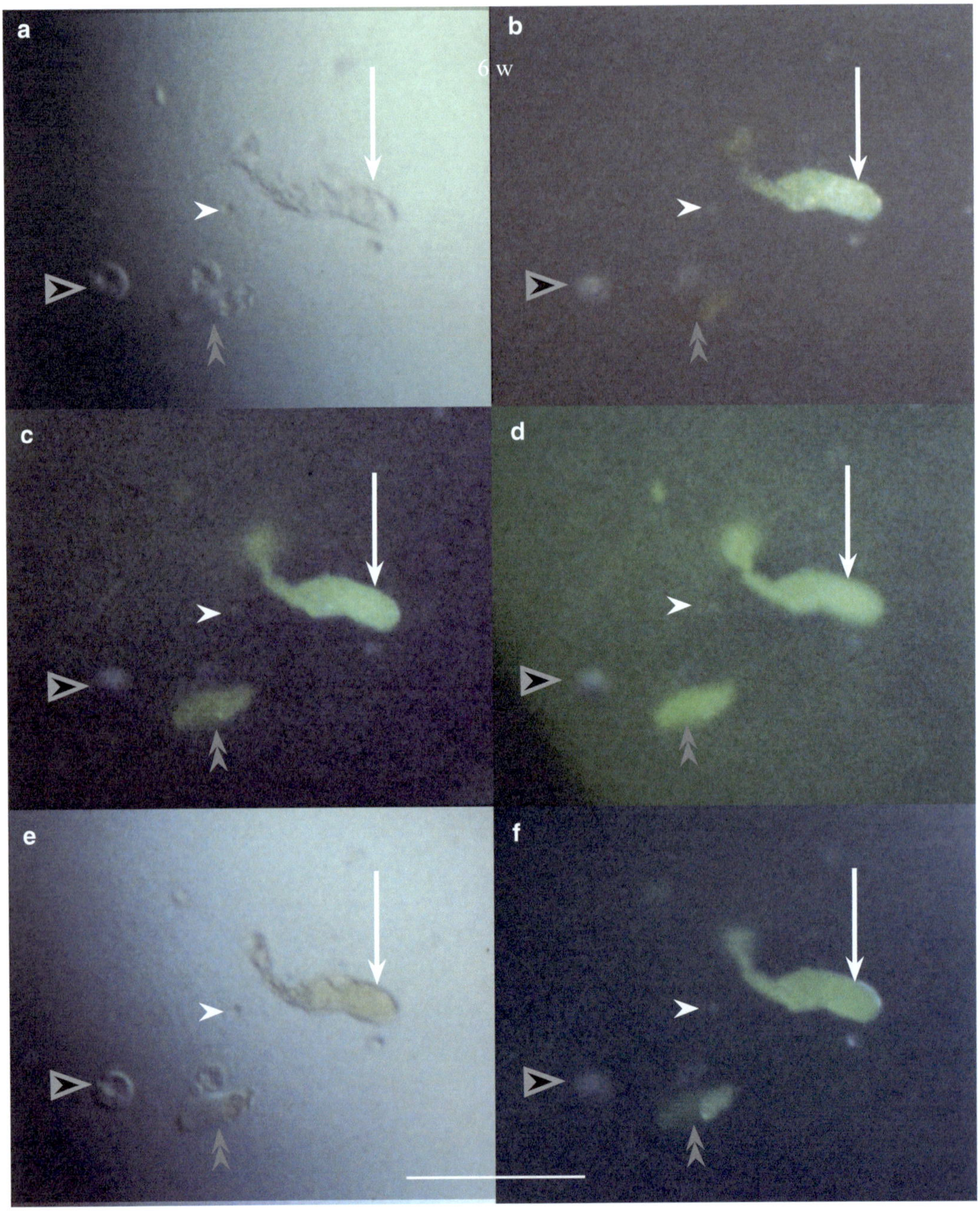

Fig. 2.15 (a) A small lesion (*arrow*), a cyst (*black arrowhead*), grouped cysts and rounded bodies (*double arrowhead*) and a rounded body (*white arrowhead*). (b–d, f) Sequential photographs of fluorescein staining. In (d), the tear film was refreshed after a blink. (e) shows the brownish hue of fluorescein in retroillumination (the markers are placed in corresponding locations)

Case 6. Recurrent Erosions or KCS Symptoms? (cont.)

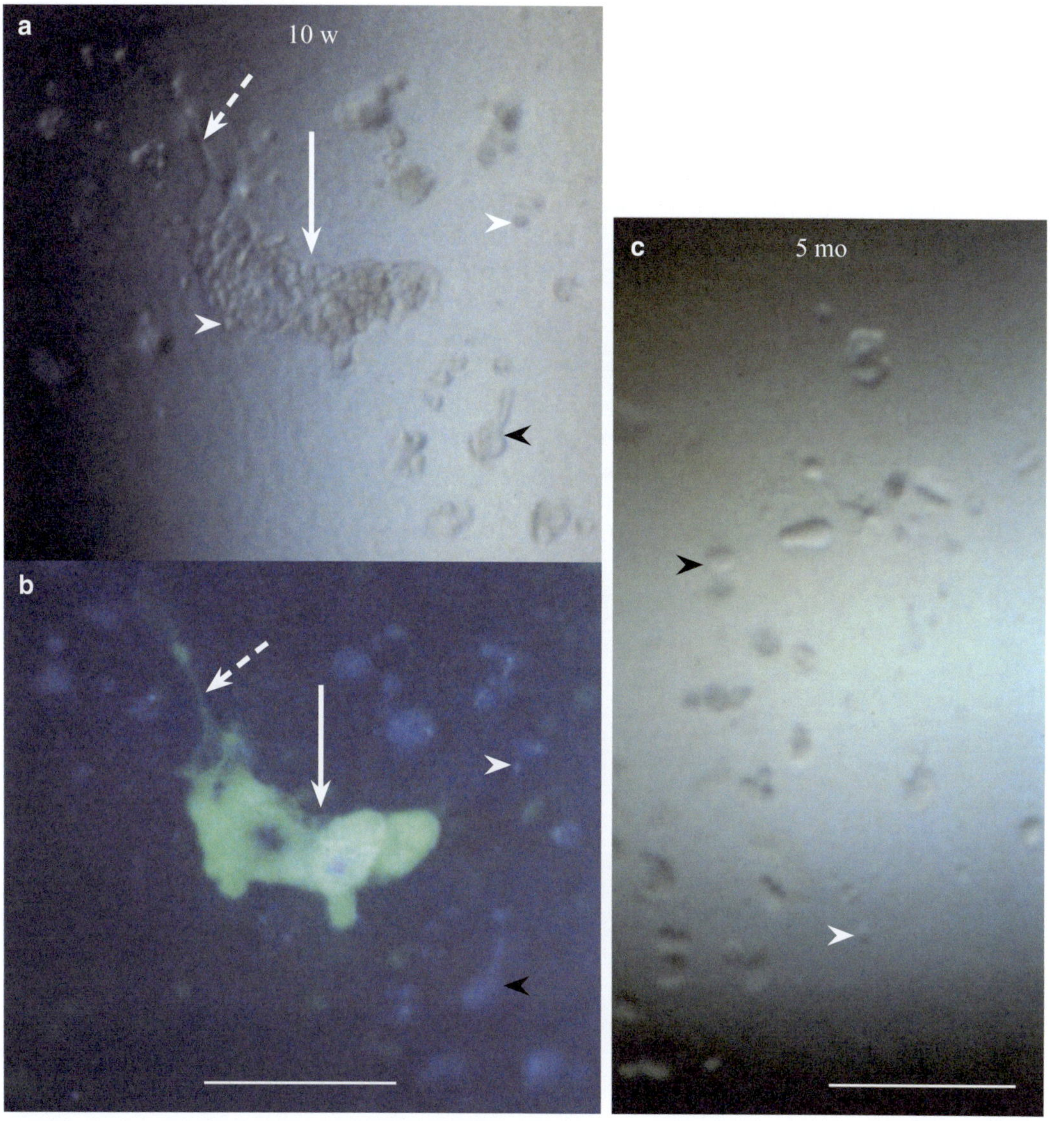

Fig. 2.16 (**a**, **b**) A lesion (*long arrows*) showing many rounded bodies (*white arrowhead*) and an epithelial line (*dashed arrows*), both (**b**) staining green with fluorescein. The surroundings show many light-reflecting rounded bodies (*white arrowheads*) and cyst-like structures (*black arrowheads*) that do not stain. (The markers are placed in corresponding locations.) (**c**) At this occasion were present several cysts (*black arrowhead*) and rounded bodies (*white arrowhead*) but no lesions

Addendum

Although symptom-free, the patient continued using NaCl ointment 5 % at night. Nine and 15 months after presentation, both corneae showed small epithelial cysts; 1 and 2 years later, when last seen, both corneae appeared normal.

Case 7. Recurrent Erosion or Herpes Simplex?

Case Report

A-48-year-old man with difficulties of verbal communication reported recurrent grittiness in the right eye on awakening for about 1 year; he had been repeatedly treated elsewhere with acyclovir because of a (presumed) virus infection. At presentation, both lower corneae showed small cysts and fingerprint lines. He was treated with NaCl ointment 5 % at bedtime, and within 1 month the pain and the cysts disappeared. Seven years later he presented again with the same symptoms, this time in both eyes. The right cornea showed small cysts and fingerprint lines and the left one additionally epithelial lesions. After 2 weeks treatment with NaCl ointment 5 % the symptoms had almost disappeared, but the corneae showed changes as before. After a further 8 weeks, the symptoms were gone; both corneae showed only cysts and fingerprint lines. When last seen, 15 months after presentation, the left cornea again showed an epithelial lesion in addition to cysts and fingerprint lines, present in both corneae; the patient's symptoms were, however, minimal and he was using the ointment only sporadically.

The photographs of the left cornea were taken during the second documented episode, at presentation (day 1) and 2 weeks, 8 weeks and 15 months later.

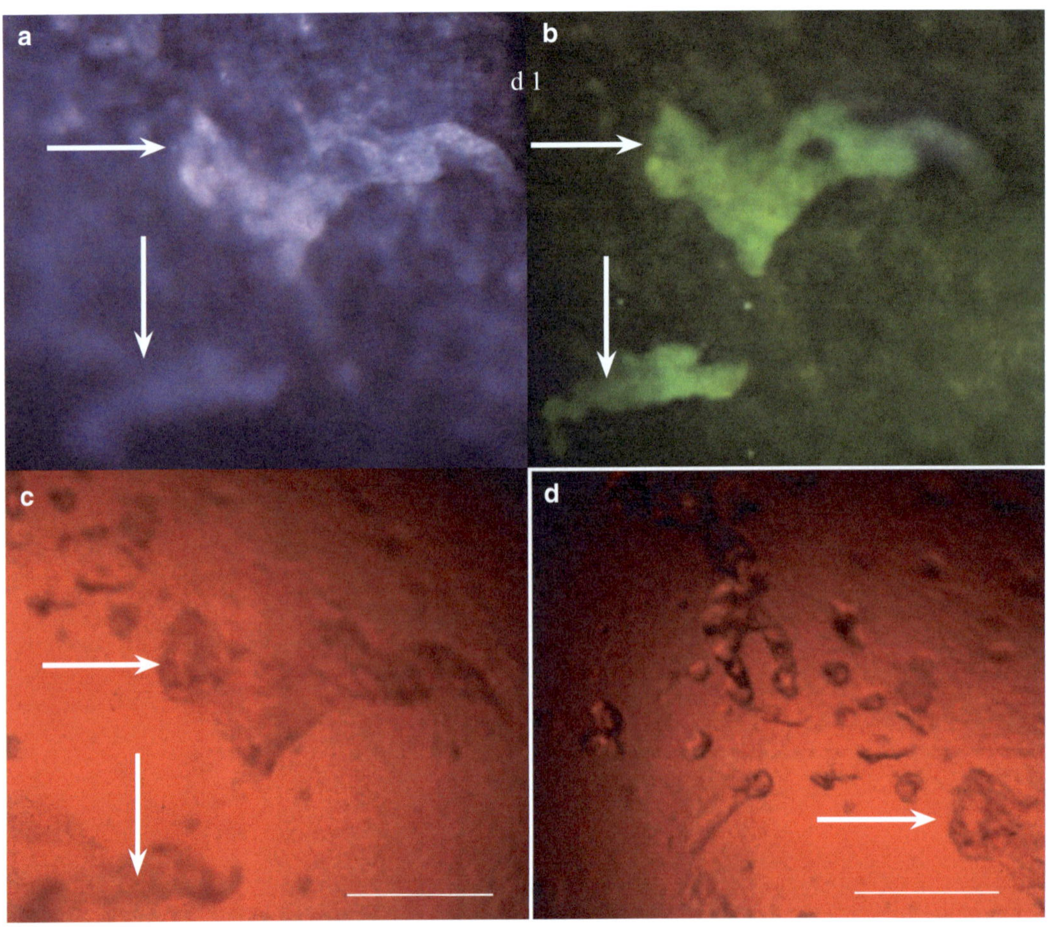

Fig. 2.17 (a) Two light-reflecting lesions (*arrows*) which (b) stain green with fluorescein; (c) the same area in retroillumination. (d) An area adjacent to the larger lesion (*arrow*) shows many cysts (*arrowhead*), individual or grouped (*double arrowhead*) (the arrows are placed in corresponding locations)

Case 7. Recurrent Erosion or Herpes Simplex? (cont.)

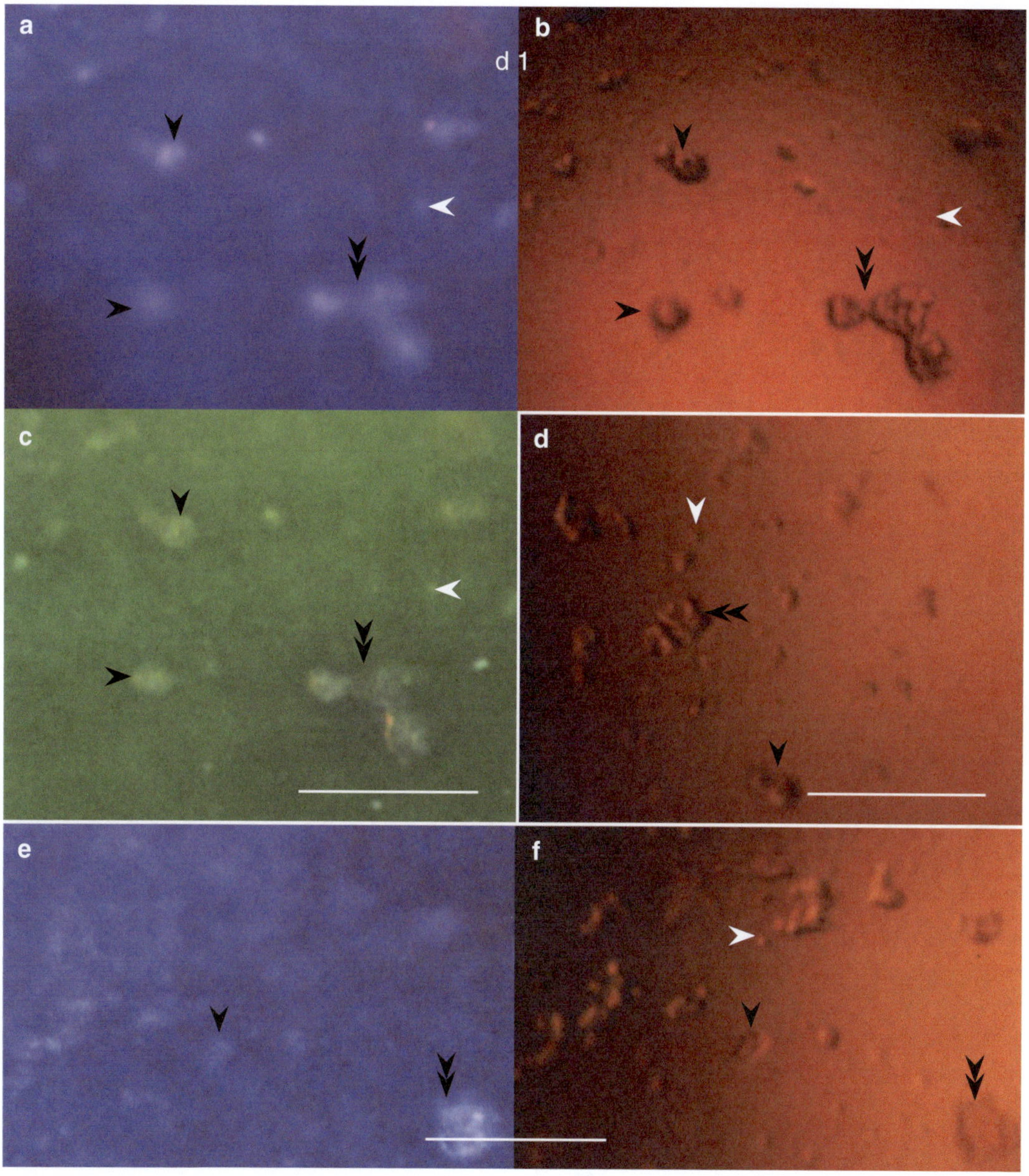

Fig. 2.18 (a–c) An area showing many (a) light-reflecting (*arrowheads*) structures which (b) in retroillumination are visible as rounded bodies (*white arrowheads*) and cysts (*black arrowheads*), partly grouped (*double arrowheads*), (c) all appearing brighter in the tear film stained green with fluorescein. (The markers are placed in corresponding locations.) (d–f) Similar changes present in different areas of the same cornea (the black arrowheads are placed in corresponding locations in each set of photographs)

Case 7. Recurrent Erosion or Herpes Simplex? (cont.)

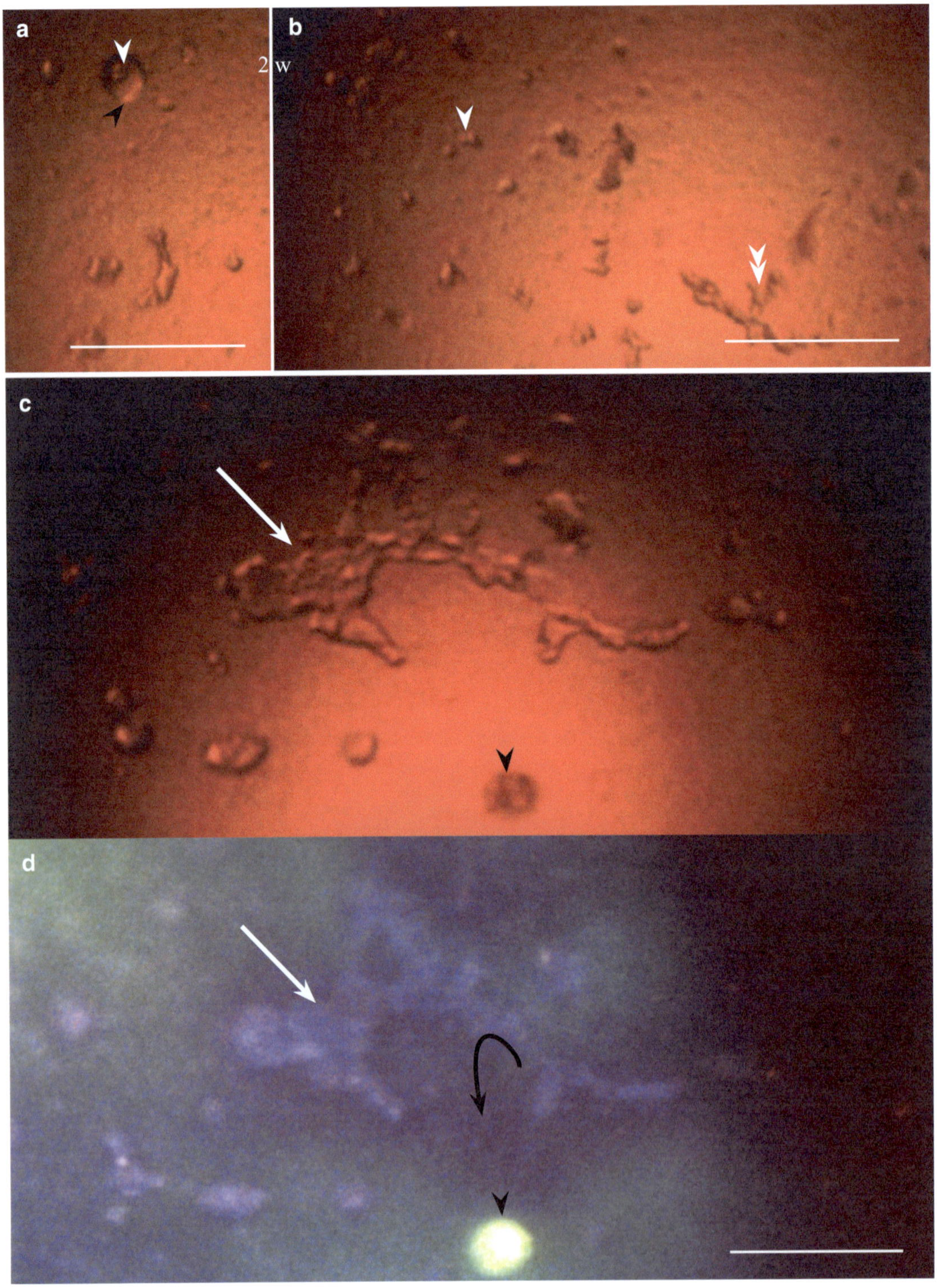

Fig. 2.19 (**a**) A cyst (*black arrowhead*) containing rounded bodies (*white arrowhead*), (**b**) rounded bodies, individual (*arrowhead*) or grouped (*double arrowhead*), and (**c**, **d**) a lesion (*white arrows*) that (**d**) does not stain but causes surface elevation (dark, *bowed arrow*). The *black arrowhead* in (**c**) indicates a cyst (**d**) the strong light reflection from which hampers the visibility of its green fluorescein staining (in (**c**, **d**), the white arrows and the arrowheads are placed in corresponding locations)

Case 7. Recurrent Erosion or Herpes Simplex? (cont.)

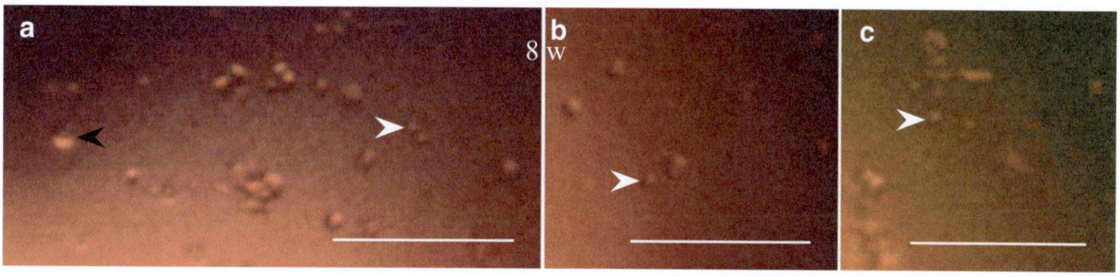

Fig. 2.20 At this occasion, the cornea showed many cysts (*black arrowheads*) and rounded bodies (*white arrowheads*)

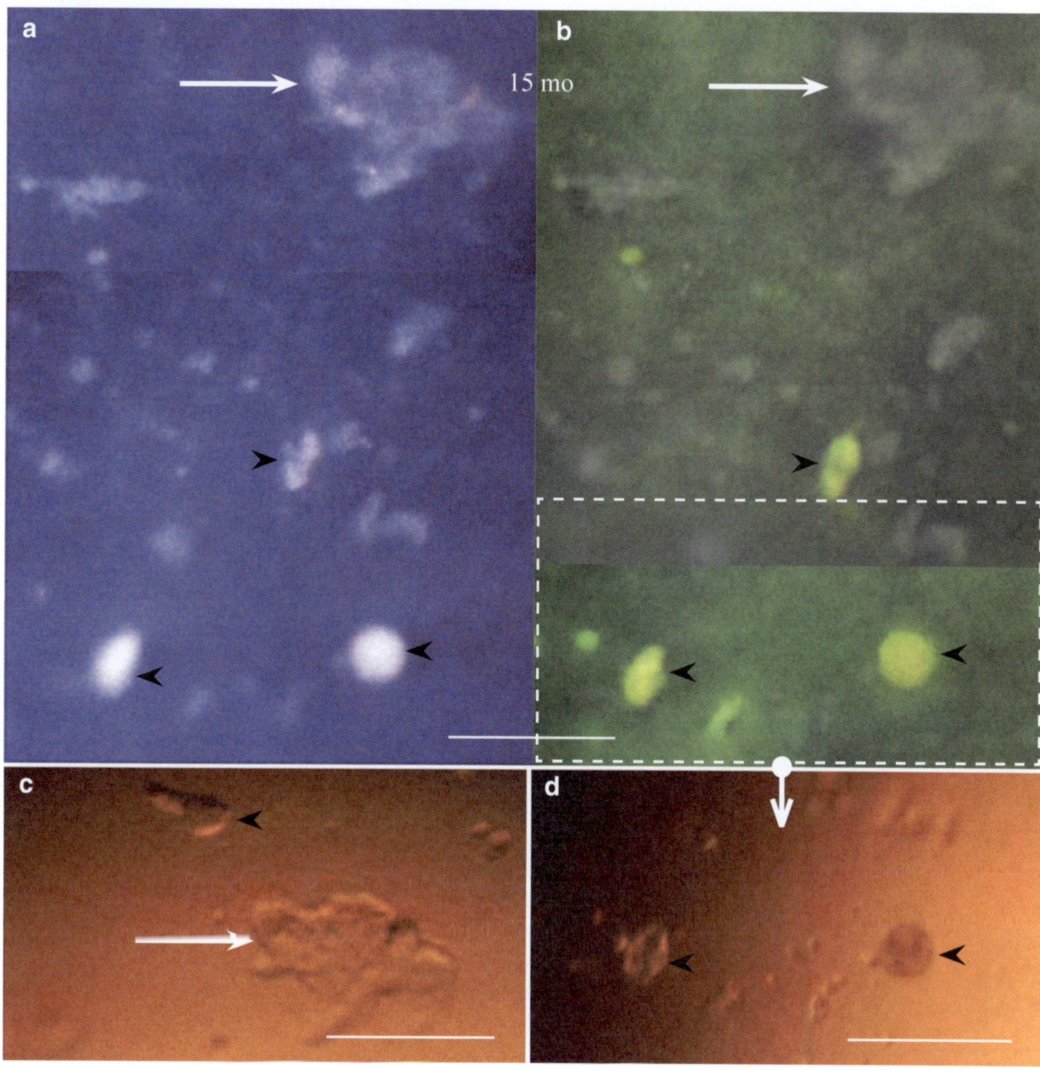

Fig. 2.21 (**a**) A light-reflecting lesion (*arrow*) and many light-reflecting structures (*arrowheads*). (**b**) With fluorescein, some cystic spaces (*arrowheads*) stain; the yellow hue of the staining is caused by the strong light reflection. (**c**) shows the lesion (*arrow*) and (**d**) cystic spaces (*arrowheads*) in the lower part of the area in retroillumination (the markers are placed in corresponding locations; (**a**, **b**) are composed photographs)

Case 8. Progress of a Lesion Occurring Within 4 h

Case Report

A 64-year-old woman with pain on awakening in both eyes for many years presented because of eye floaters. Both corneae showed epithelial derangement compatible with recurrent erosion syndrome and fingerprint lines. Goldman's applanation tonometry was performed.

The photographs of the right cornea were taken immediately after the tonometry and 4 h later.

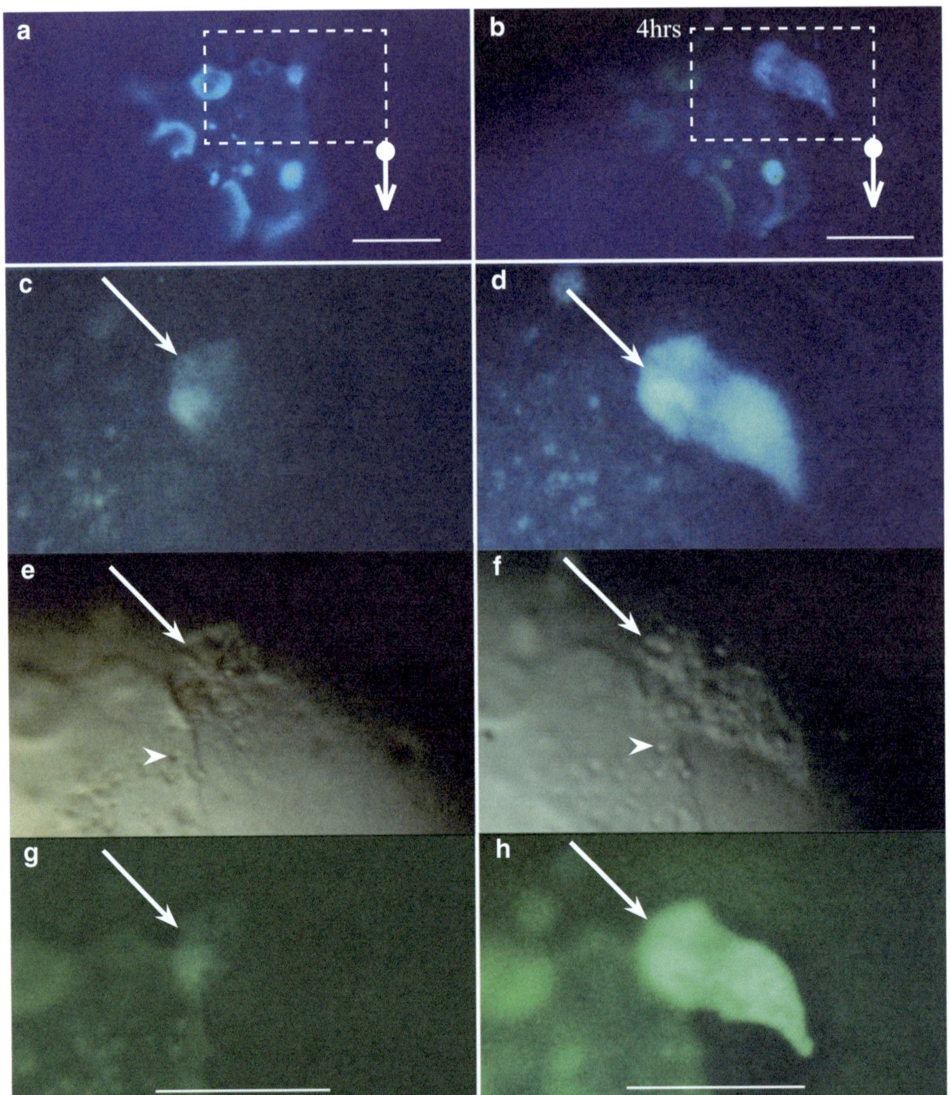

Fig. 2.22 (a, b) Survey of a lesion (a) immediately after tonometry and (b) 4 h later (fluorescein and blue filter). *Left column*: the area in *frame* in (a); *right column*: the area in *frame* in (b), captured (c, d) in focal illumination, (e, f) in retroillumination and (g, h) after the application of fluorescein. A part of the lesion (*arrows*) has enlarged. In (e, f) some rounded bodies (*arrowheads*) are visible that have remained in place (the markers are placed in corresponding locations)

Case 9. Regress of a Lesion (1)

Case Report

A 33-year-old woman with pain on awakening in the left eye for 5 years (after having scratched the cornea with a fingernail). At presentation, the right cornea appeared normal. The left eye was white, but the lower nasal cornea showed a large epithelial lesion. She started NaCl ointment 5 % at bedtime on the following day. Six weeks later, she was symptom-free; the left cornea showed a few cysts in the area of the previous lesion. After a symptom-free year, the same problem returned but disappeared rapidly with NaCl ointment 5 %. At that occasion, the right cornea showed no changes and the left one a few cysts. After 5 symptom-free years she presented again, this time with symptoms in the right eye, but both corneae appeared normal. Another 5 symptom-free years passed before she had a new, short, episode in the right eye. The right cornea showed discrete fingerprint lines and a few cysts, and the left one appeared normal. She reported that her father had similar problems after a trivial injury.

The photographs of the left cornea were taken at presentation (day 1), on 3 consecutive days and 6 weeks and 1 year later.

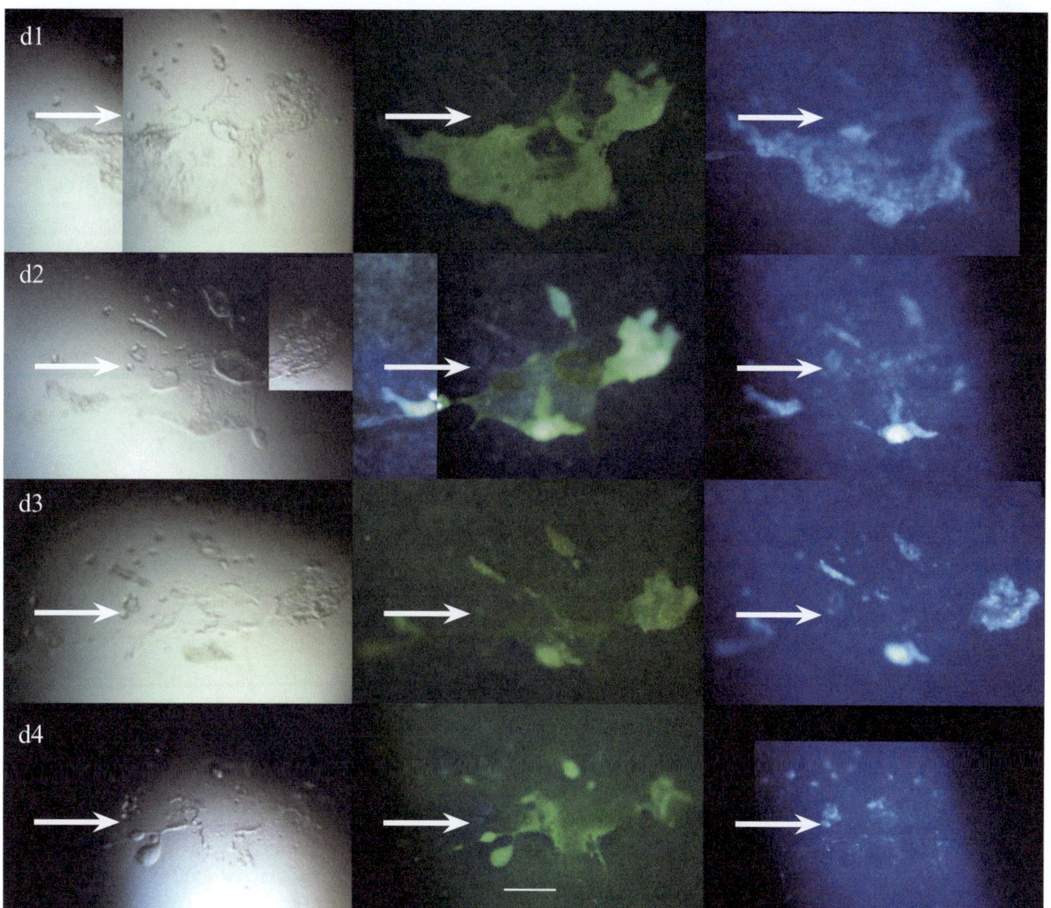

Fig. 2.23 Survey of a lesion captured on 4 consecutive days in retroillumination (*left column*), after the application of fluorescein (*central column*) and in focal illumination (*right column*). For orientation, the *arrows* indicate a structure identifiable at all the four occasions. For details see Figs. 2.24, 2.25 and 2.26 (overleaf) (*d1* left column and *d2* left and central columns, are composed photographs)

Case 9. Regress of a Lesion (1) (cont.)

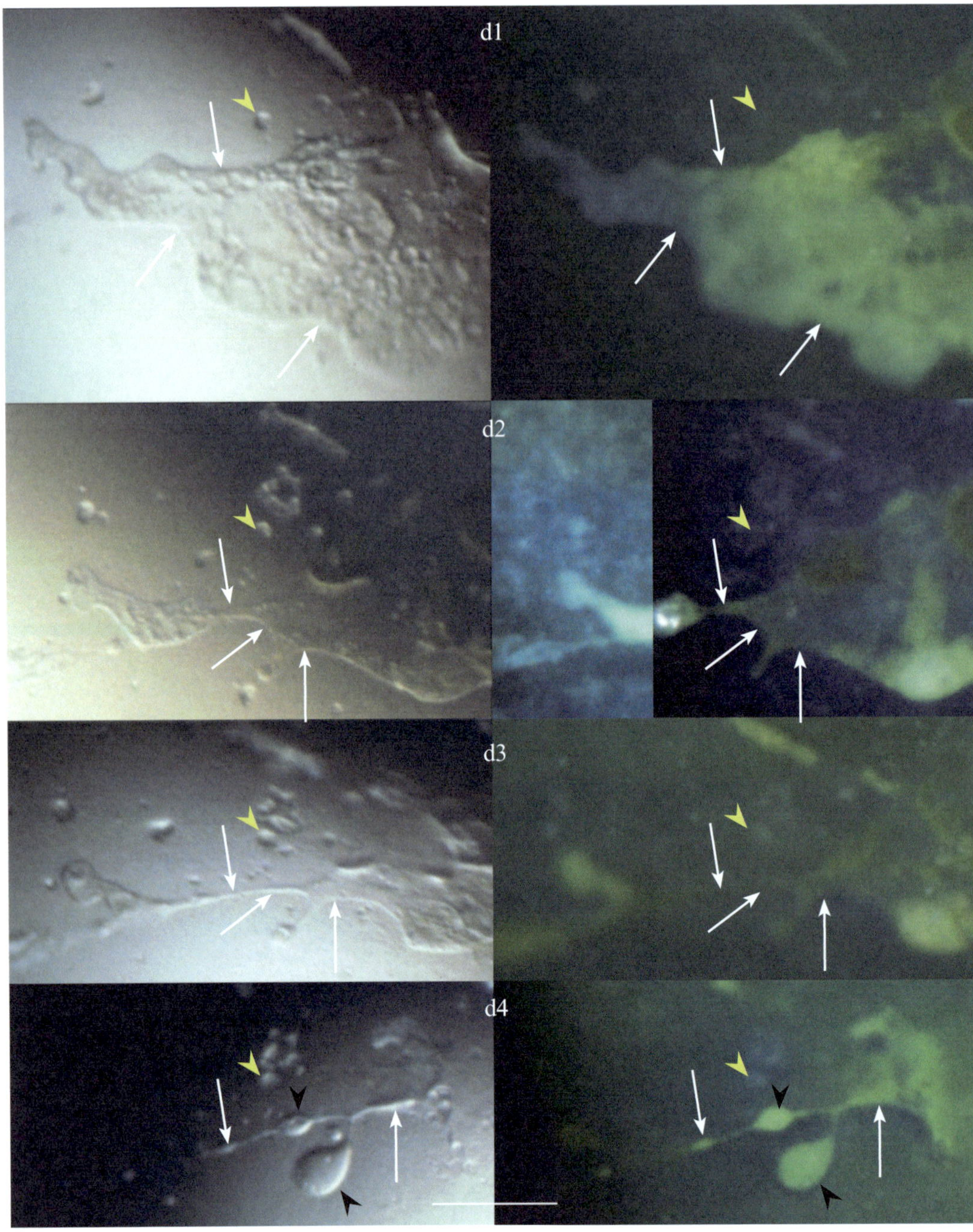

Fig. 2.24 Left part of the lesion shown in Fig. 2.23, captured on 4 consecutive days (d1–d4), in retroillumination (*left column*) and with fluorescein (*right column*). The *yellow arrowheads* indicate a structure identifiable at all the four occasions and the *arrows* advancing edges of epithelial sheets; in places, just before merging, the opposite edges appear as thin lines (d3 and d4). In some areas of contact between the edges develop cysts (d4, *black arrowheads*). Throughout, fluorescein shows various shades of green (the markers are placed in corresponding locations in each pair of photographs; *d2* (right column) is a composed photograph)

Case 9. Regress of a Lesion (1) (cont.)

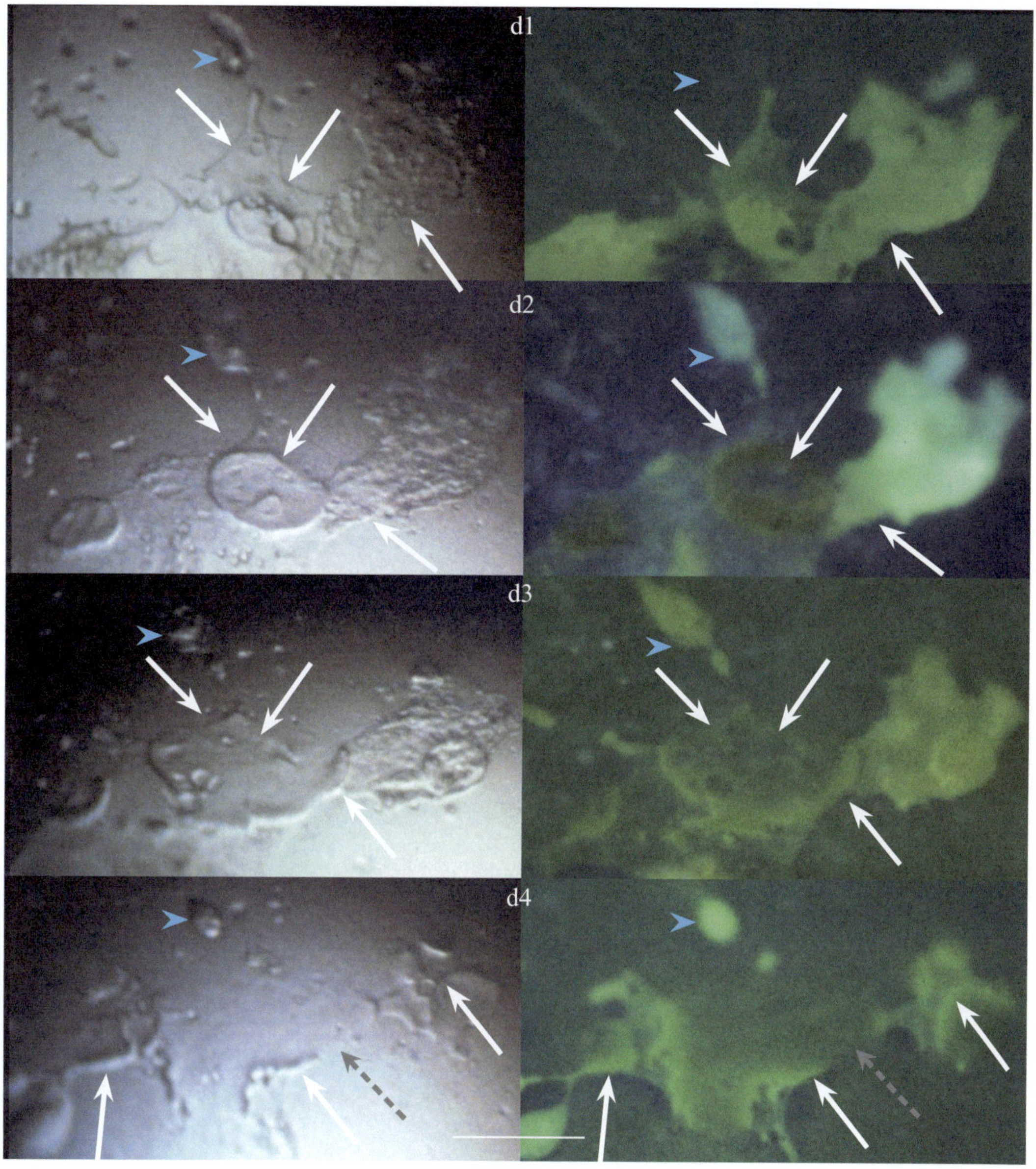

Fig. 2.25 Right upper part of the lesion shown in Fig. 2.23, captured on 4 consecutive days (d1–d4) in retroillumination (*left column*) and with fluorescein (*right column*). For orientation, the *blue arrowheads* indicate an approximately comparable location in all photographs and the *arrows* (d1 and d2) edges of epithelial sheets. Note changes occurring in the centre of the lesion between the three *arrows* (the markers are placed in corresponding locations in d1–d3). On d4 remaining curved edges (*plain arrows*) of epithelial sheets are visible; the *dashed arrow* indicates an area in which the opposite edges seem to have merged. Fluorescein staining shows various shades of green; elevated structures appear dark (the markers are placed in corresponding locations in each pair of photographs)

Case 9. Regress of a Lesion (1) (cont.)

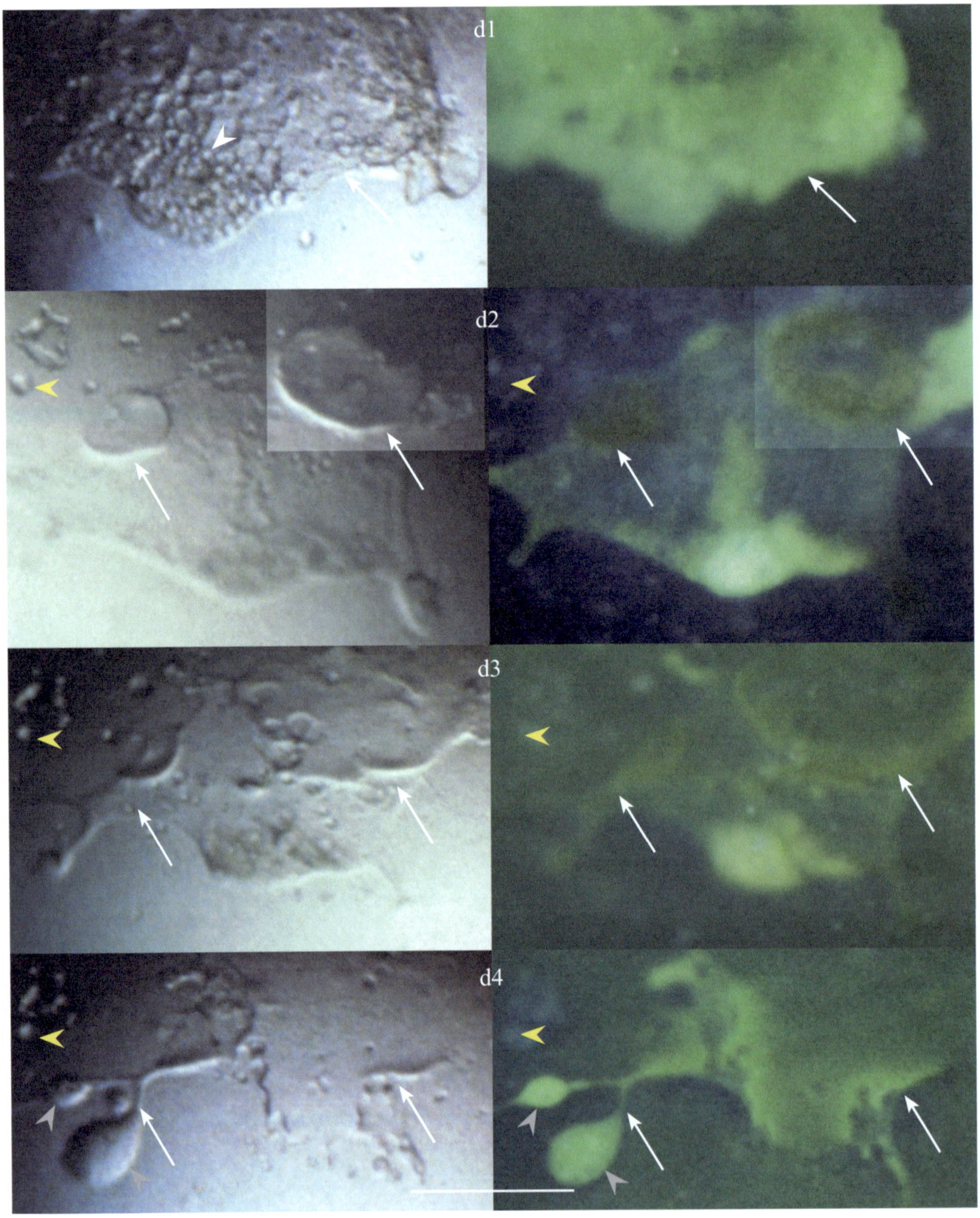

Fig. 2.26 Lower part of the lesion shown in Fig. 2.23, captured on 4 consecutive days (d1–d4) in retroillumination (*left column*) and with fluorescein (*right column*). On (d1), the lesion (*arrow*) contains many rounded bodies (*arrowhead*). On (d2–d4), for orientation, the *yellow arrowheads* indicate a structure identifiable in all the images, and the *arrows* are placed at the same distance from it. The edges advance toward each other and the disturbed epithelium between them is successively eliminated. In d4 a cystic expansion and a pear-shaped cyst (*grey arrowheads*) developing in the path of merging epithelial sheets are visible

Case 9. Regress of a Lesion (1) (cont.)

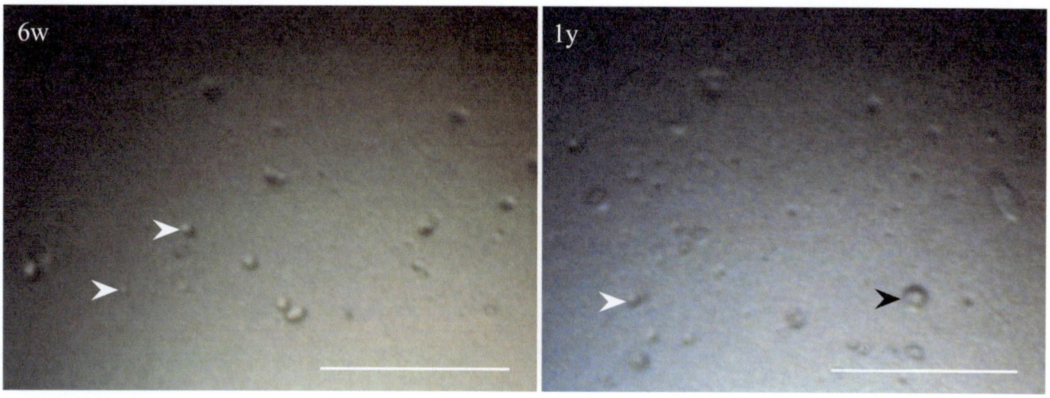

Fig. 2.27 Rounded bodies (*white arrowheads*) and small cysts containing rounded bodies (*black arrowhead*)

Case 10. Regress of a Lesion (2)

Case Report

A 30-year-old woman had touched her right eye with the blunt end of a pen. Twenty-four hours later, at presentation, the right eye was injected and the lower part of the cornea showed a large area of edematous epithelium containing a small disruption. The lower part of the left cornea showed a few cysts. She was treated with antibiotic ointment. On the following day, NaCl ointment 5 % at bedtime was added. Two weeks after the accident she had minimal symptoms, and after a further 4 weeks she was symptom-free; the right cornea showed small epithelial cysts. She stopped the ointment. When last seen, 6 months after the accident, she was symptom-free; both corneae showed small epithelial cysts.

The photographs of the right cornea were taken on day 1 (at presentation), day 2, day 3 and day 4 and 2 weeks, 6 weeks and 6 months after the accident.

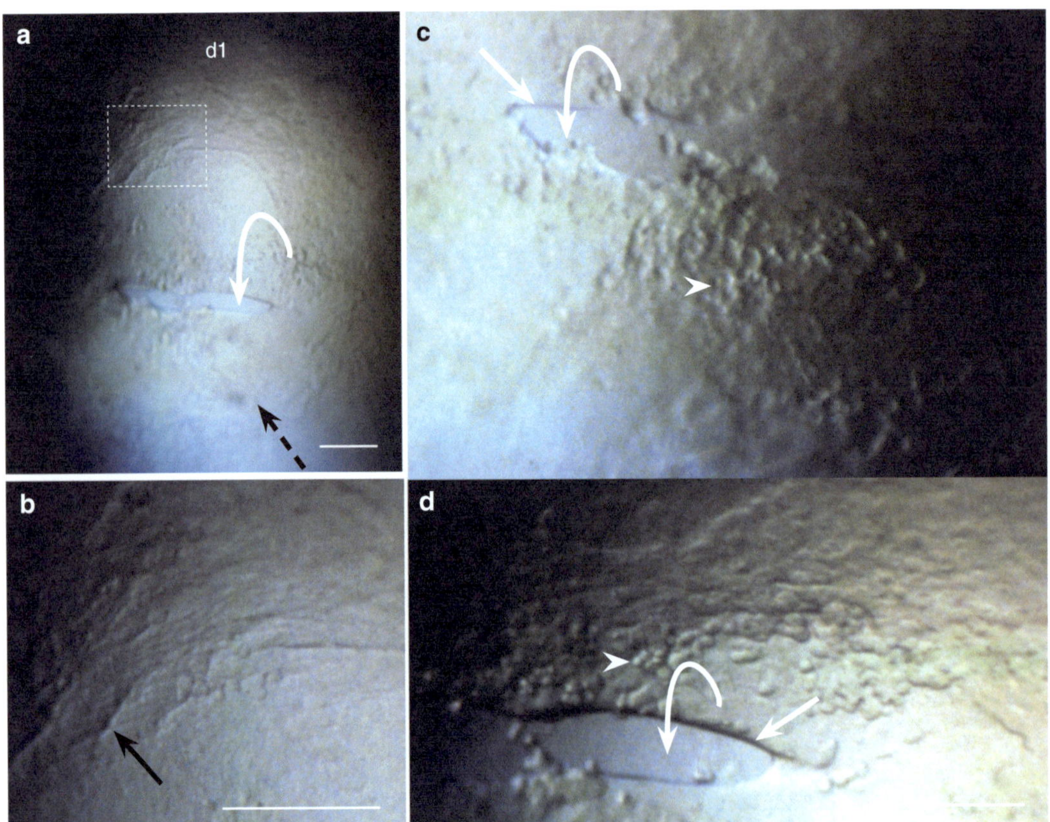

Fig. 2.28 (**a**) Lower part of a large edematous area; the *dashed arrow* indicates its lower edge, the *frame* the area shown at higher magnification in (**b**) and the *bowed arrow* the area shown in (**c**, **d**). (**b**) The epithelium shows lines (*arrow*) suggestive of epithelial wrinkles. (**c**) shows the left and (**d**) the right part of an oval area (*plain arrows*), the bottom of which (*bowed arrows*) appears smooth. The surroundings show many rounded bodies (*arrowheads*)

Case 10. Regress of a Lesion (2) (cont.)

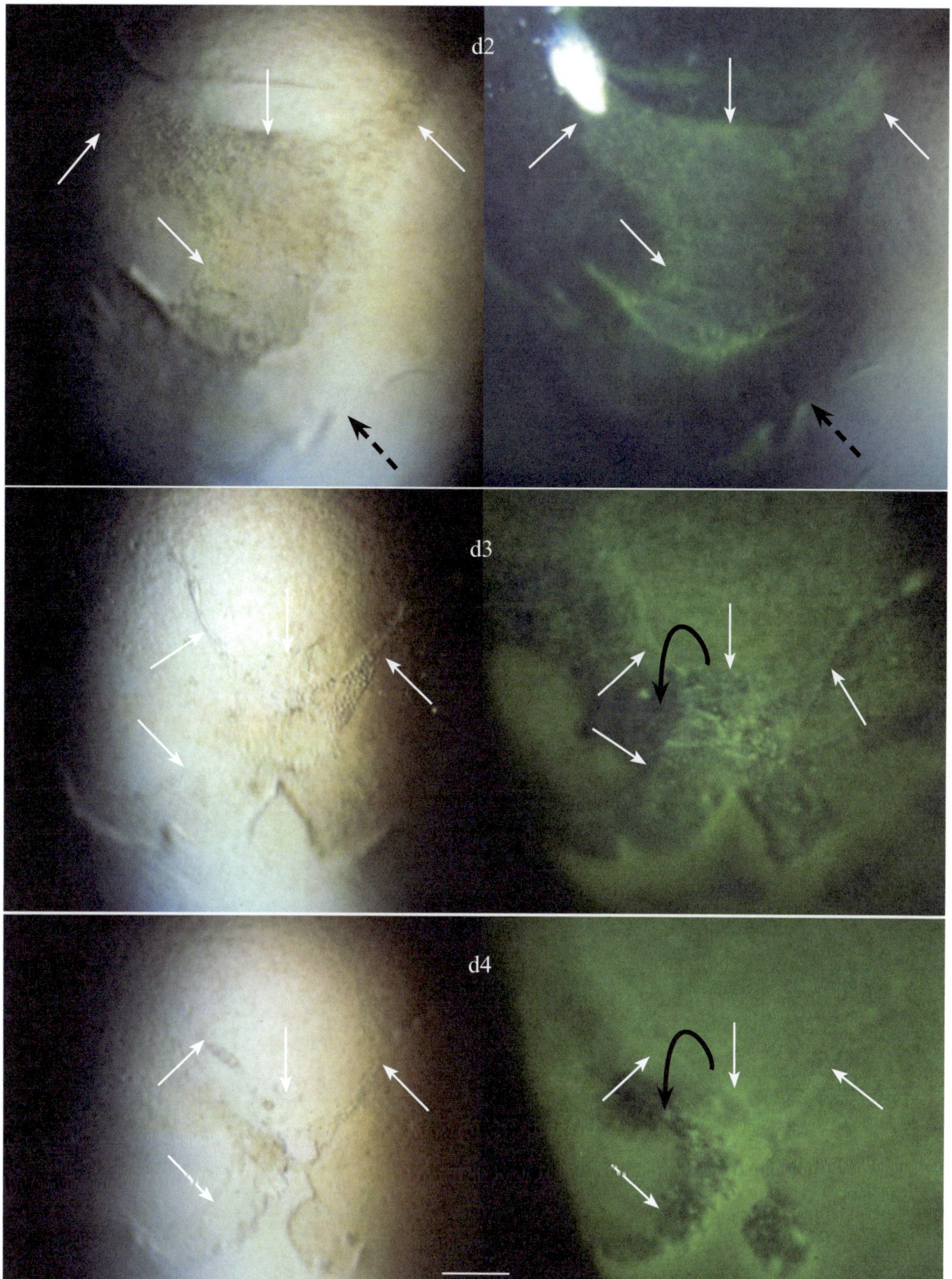

Fig. 2.29 Survey of changes occurring in the area shown in Fig. 2.28 on 3 consecutive days. In (d2), the *dashed arrows* indicate lower edge of the edematous area containing the lesion; lacking reference points, the *white arrows* indicating the lesion's edges serve orientation purposes for comparison with d3 and d4. Fluorescein shows (d2) no diffusion into the surroundings. In (d3–d4) surface elevations (dark, *bowed arrows*) in the green stained tear film are visible. (The markers are placed in corresponding locations in each pair of photographs.) For details see Figs. 2.30, 2.31, 2.32, 2.33 and 2.34

Case 10. Regress of a Lesion (2) (cont.)

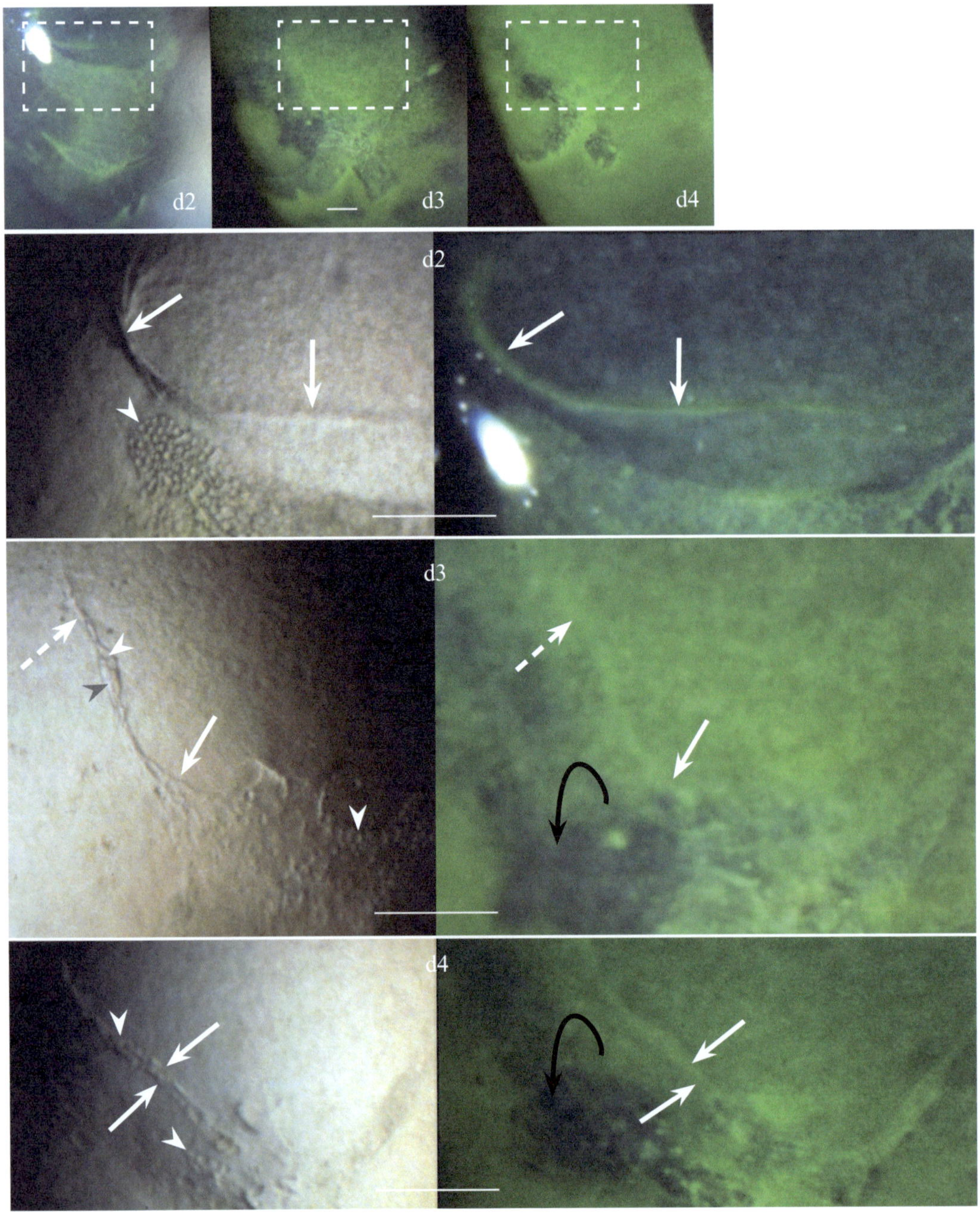

Fig. 2.30 Left upper part (*upper row, in frames*) of the lesion on 3 consecutive days captured before (*left column*) and after fluorescein staining (*right column*). It shows (d2) a curved edge (*arrows*) of an epithelial sheet and many rounded bodies (*arrowhead*); (d3) a curved edge of an epithelial sheet (*plain arrows*), an epithelial line (*dashed arrows*) with an expansion (*grey arrowhead*) and rounded bodies within the expansion and within the lesion (*white arrowheads*); and (d4) a cleft between the edges of epithelial sheets (*arrows*); the cleft and the lesion contain rounded bodies (*arrowheads*). In (d3–d4), fluorescein is still present in the tear film in which the line and the cleft (*straight arrows*) appear brighter and surface elevations (*bowed arrows*) dark (the white arrows are placed in corresponding locations in each pair of photographs)

Case 10. Regress of a Lesion (2) (cont.)

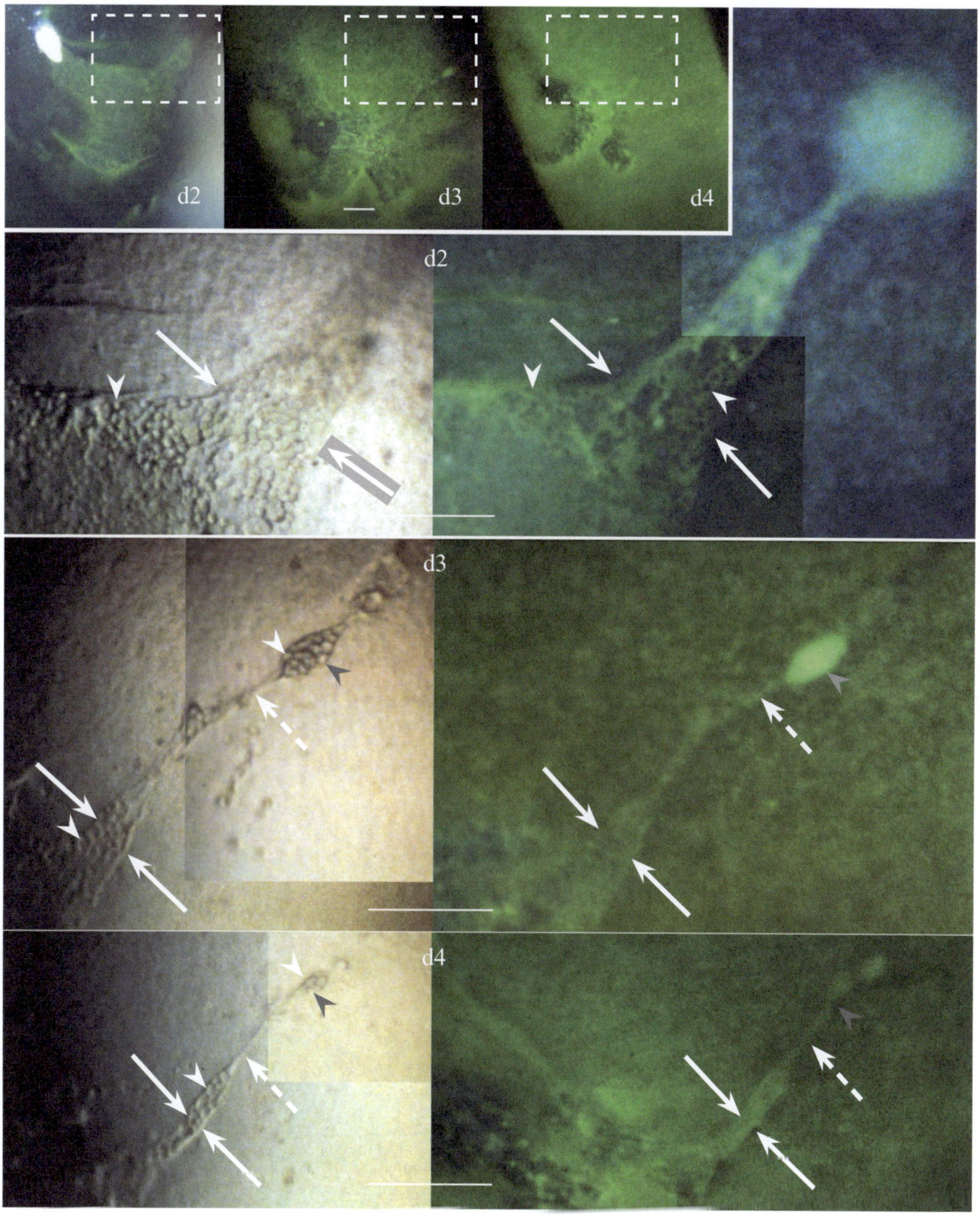

Fig. 2.31 Right upper part (*upper row*, *in frames*) of the lesion on 3 consecutive days captured before (*left column*) and after fluorescein staining (*right column*). It shows (d2) curved edges (*arrows*) of epithelial sheets and rounded bodies (*arrowheads*) and (d3–d4) edges of epithelial sheets (*plain arrows*), epithelial lines (*dashed arrows*), expansions (*grey arrowheads*) and rounded bodies (*white arrowheads*) within them and between the edges. With fluorescein, the rounded bodies appear as rounded dark dots (d2, *arrowheads*); in (d3–d4), the spaces between the edges (*arrows*) appear brighter than the green stained tear film and the cystic expansion (*arrowhead*) brilliantly green (the arrows and the grey arrowheads are placed in corresponding locations in each pair of photographs; *d2* right column and *d3–d4* left column are composed photographs)

Case 10. Regress of a Lesion (2) (cont.)

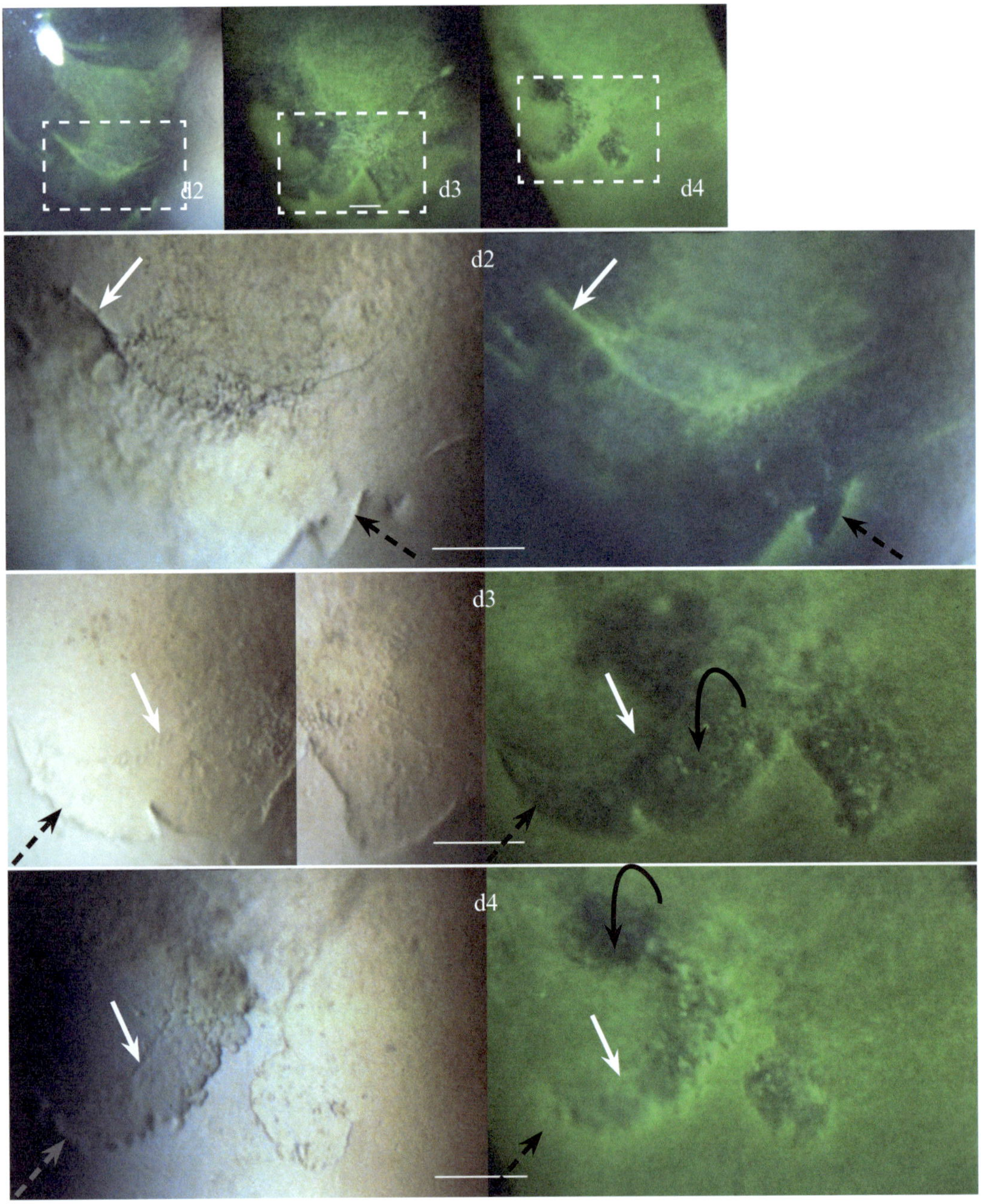

Fig. 2.32 Lower part (*upper row*, *in frames*) of the lesion captured on 3 consecutive days before (*left column*) and after fluorescein staining (*right column*). (d2) The edge of the lesion (*dashed arrows*) seems elevated (dark); the *white arrows* indicate a curved edge of a fusiform area within the lesion (cf. Fig. 2.33) delineated by green stained lines. In (d3–d4), the *dashed arrows* indicate the left lower edge of the lesion and the *white arrows* a curved edge of an epithelial sheet. With fluorescein, the elevated parts of the lesion (*bowed arrows*) appear dark (the straight arrows are placed in corresponding locations in each pair of photographs; *d3*, left column, is a composed photograph)

Case 10. Regress of a Lesion (2) (cont.)

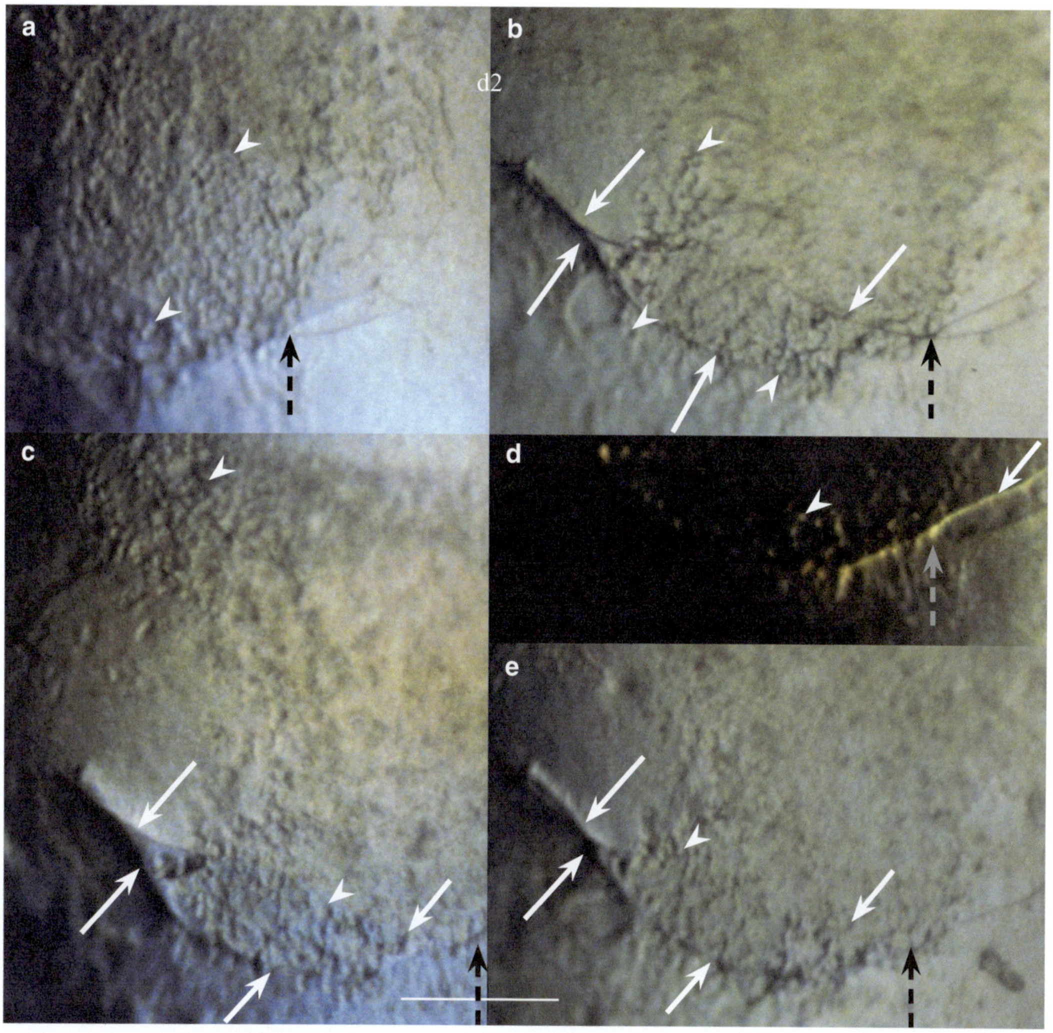

Fig. 2.33 The fusiform area shown in Fig. 2.32 (d2) and its surroundings. For orientation, the *dashed arrows* indicate the same location in all photographs. (**a–d**) Both the area and its surroundings show rounded bodies (*arrowheads*). The *white arrows* indicate curved lines delineating the area; they seem to cross each other (*dashed arrows*); in (**b**, **c**, **e**) is visible a cleft between the smooth edges (*left white arrows*) and in (**d**) is visible an edge (*white arrow*) that in (**a**, **b**, **e**) appears as a line to the right of the *dashed arrow*. (The white arrows in **b**, **c** and **e** are placed in corresponding locations.) See also Fig. 1.46

Case 10. Regress of a Lesion (2) (cont.)

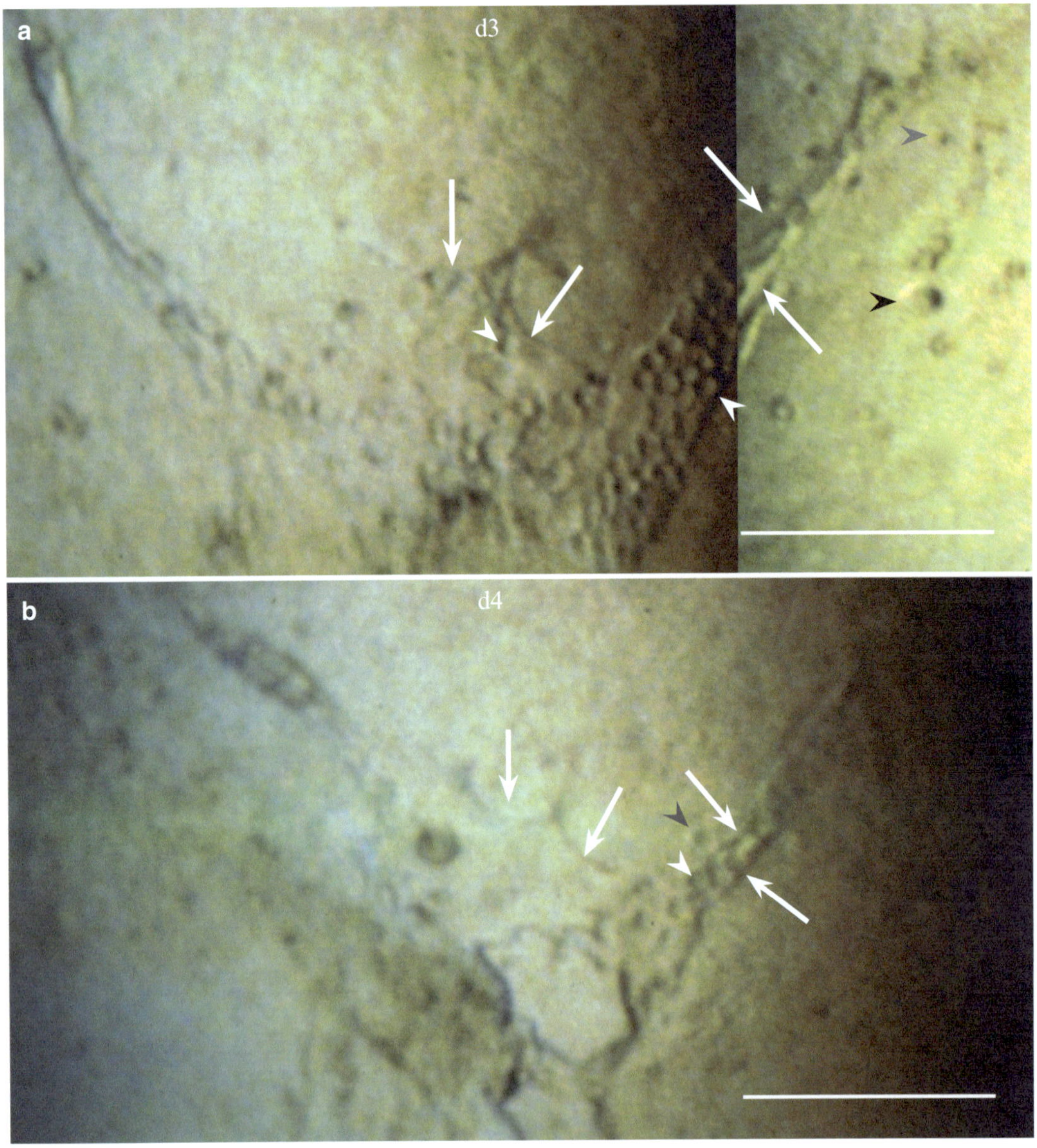

Fig. 2.34 (a, b) Close view of rounded bodies (*white arrowheads*) situated in front of or between variously curved edges (*arrows*) of epithelial sheets. Reference points are lacking but it is clearly visible that the edges to the right are approaching each other; some rounded bodies (*grey arrowheads*) and/or cysts (*black arrowhead*) are visible also peripherally from the edges. ((a) is a composed photograph)

Comment

The rounded bodies within the lesion shown in Figs. 2.28, 2.29, 2.30, 2.31, 2.32, 2.33 and 2.34 probably represent swollen epithelial cells.

Case 10. Regress of a Lesion (2) (cont.)

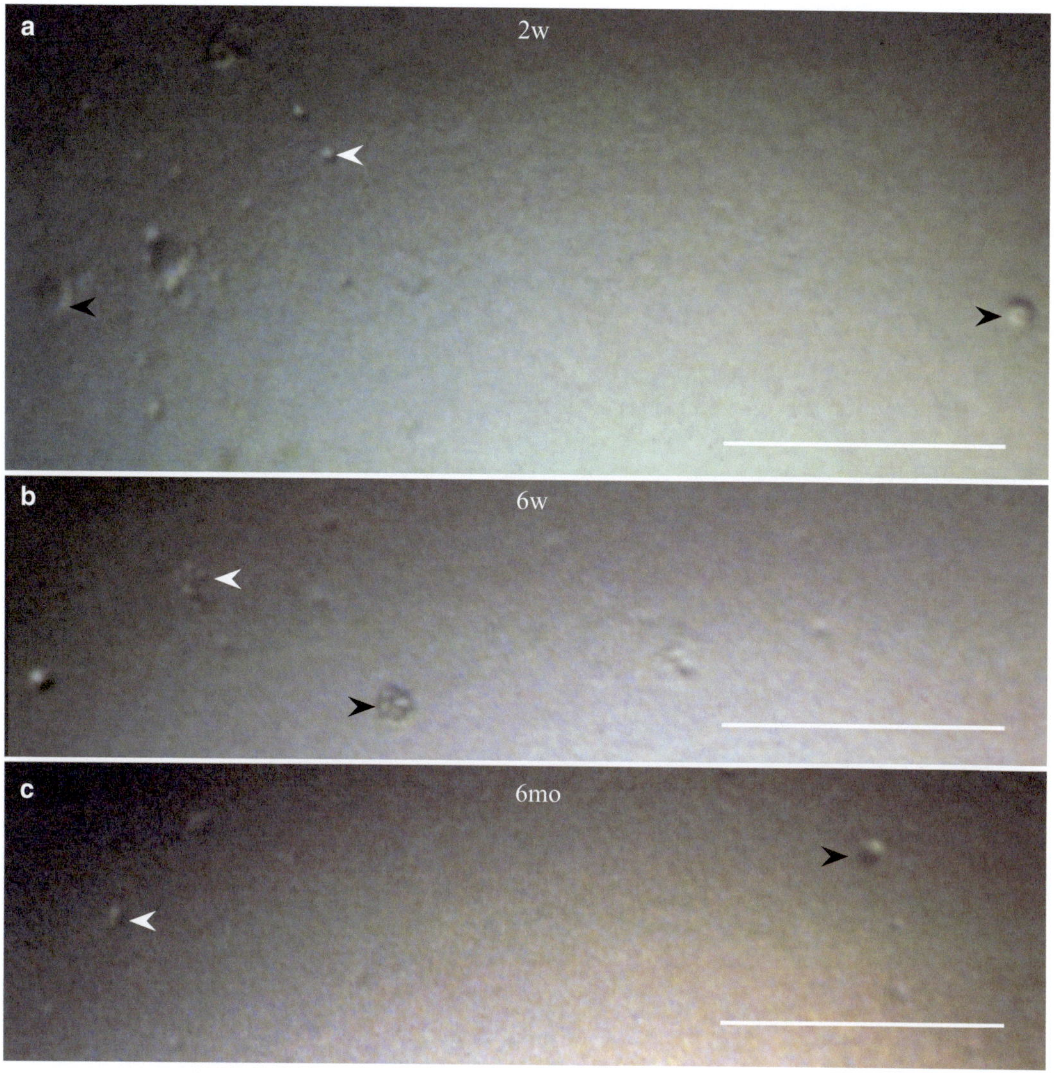

Fig. 2.35 (**a**–**c**) The previously disturbed area shows rounded bodies (*white arrowheads*) and cysts (*black arrowheads*) of which some contain rounded bodies ((**b**), *black arrowhead*)

Case 11. Surface Erosion in Predisposed Cornea (1)

Case Report

A 46-year-old man presented 2 days after having injured his right eye with his fingernail. The cornea showed an epithelial erosion. After that, he often had pain on awakening. Six months later, he presented again about 40 h after a new, similar injury. The right eye showed a large corneal erosion situated within a large edematous epithelial area engaging the lower part of the cornea, stromal edema, fine endothelial deposits, flare in the anterior chamber and miosis. The left cornea showed small cysts and fingerprint lines.

The photographs of the right cornea were taken after the second injury.

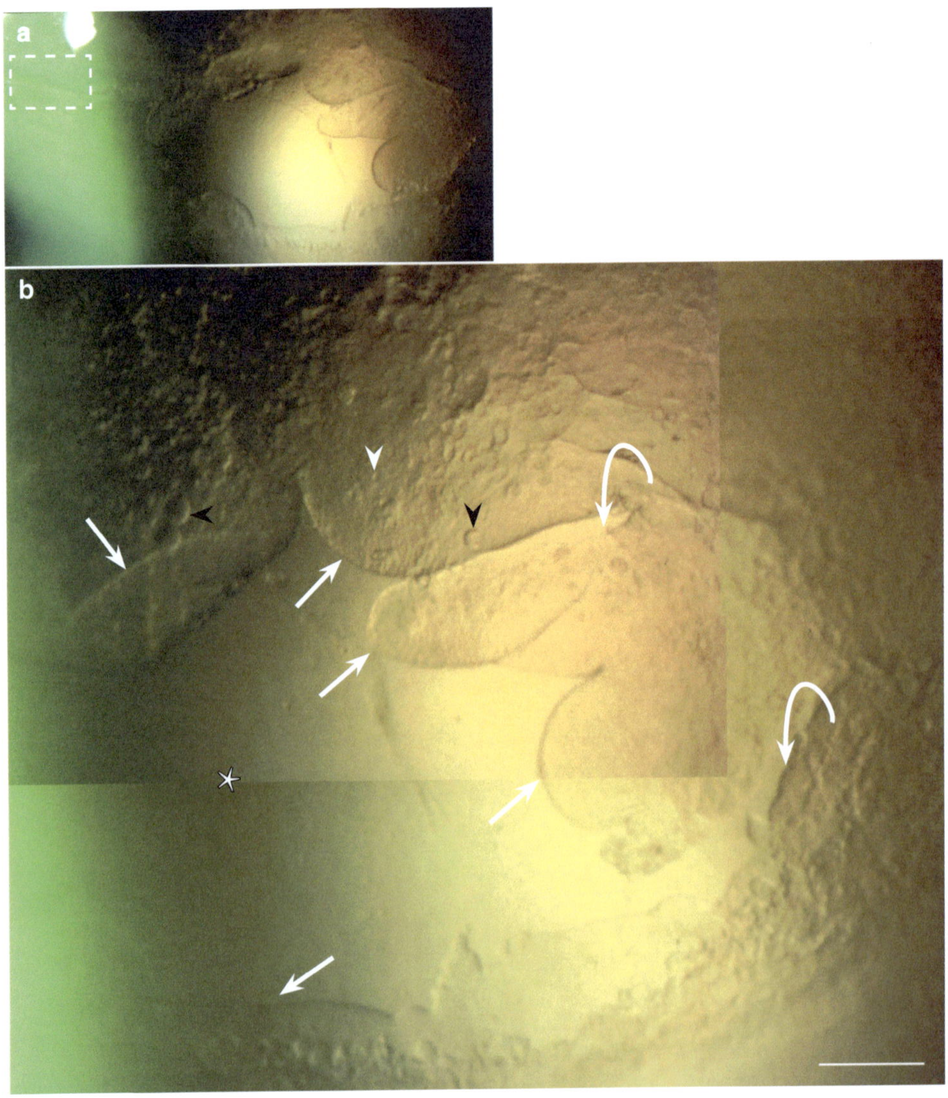

Fig. 2.36 (a) Survey of a large epithelial erosion 40 h after injury. (b) In the right part of the erosion curved edges of epithelial sheets (*straight arrows*) are visible, of which the left upper one (*straight arrow*) seems everted; the *bowed arrows* indicate spaces between the sheets, the *black arrowheads* cysts, the *white arrowheads* rounded bodies and the *asterisk* the bottom of the erosion. The area in *frame* is shown in Fig. 2.37b, c. ((b) is a composed photograph)

Case 11. Surface Erosion in Predisposed Cornea (1) (cont.)

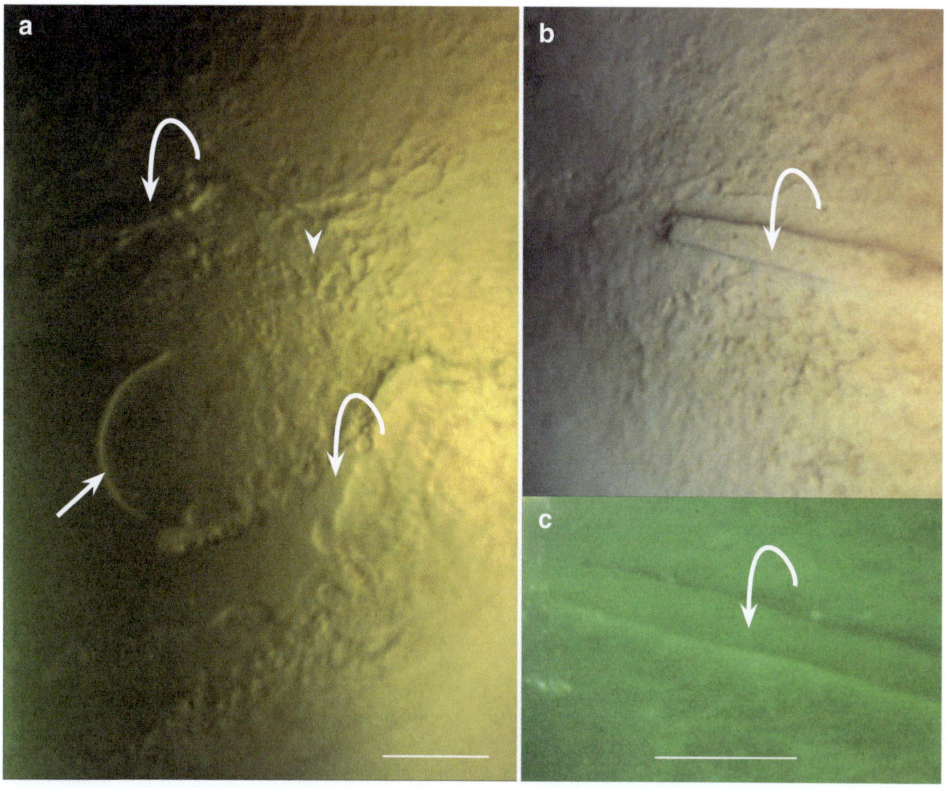

Fig. 2.37 (a) Right part of the erosion shown in Fig. 2.36 in a slightly different illumination. The curved edge of the epithelial sheet (*straight arrow*) is light-reflecting. The *bowed arrows* indicate spaces between the edges and the *arrowhead* one of the many rounded bodies present in the area. (The arrows are placed in locations corresponding to Fig. 2.36b). (b, c) In the *left part* of the erosion shown in Fig. 2.36a an epithelial depression (*bowed arrows*) is visible; (c) the green fluorescein staining implies an extensive diffusion into the tissues

Comment

The rounded bodies in Figs. 2.36b and 2.37a probably represent swollen epithelial cells.

Case 12. Surface Erosions in Predisposed Cornea (2)

Case Report

A 39-year-old man with diabetes had a large corneal erosion in the right eye caused by a dog's paw, and after that occasional episodes of grittiness in the right eye on awakening. One year later, the cornea showed a few small cysts. The left eye had little residual vision after vitrectomy complicated by endophthalmitis. After a further 8 months, he presented with pain in the right eye, this time injured by a small tree branch about 24 h previously. The eye showed a large central corneal epithelial erosion with rugged edges, fine endothelial deposits, flare in the anterior chamber and miosis. He was treated with topical antibiotics and cycloplegics. Two days later, the erosion was considerably smaller. NaCl 5 % ointment at bedtime was added, and after a further 4 days all that remained was a small area of surface irregularities.

Eleven months later, he presented again with pain and redness in the right eye, this time spontaneous. The central and paracentral lower cornea showed a large area of surface derangement. He was treated with NaCl ointment 5 % at bedtime and a week later only a few small cysts were left.

The photographs of the erosion were taken on day 1 (at presentation), day 3 and day 7, and of the spontaneous episode occurring 11 months later on day 1 (at presentation) and after 1 week.

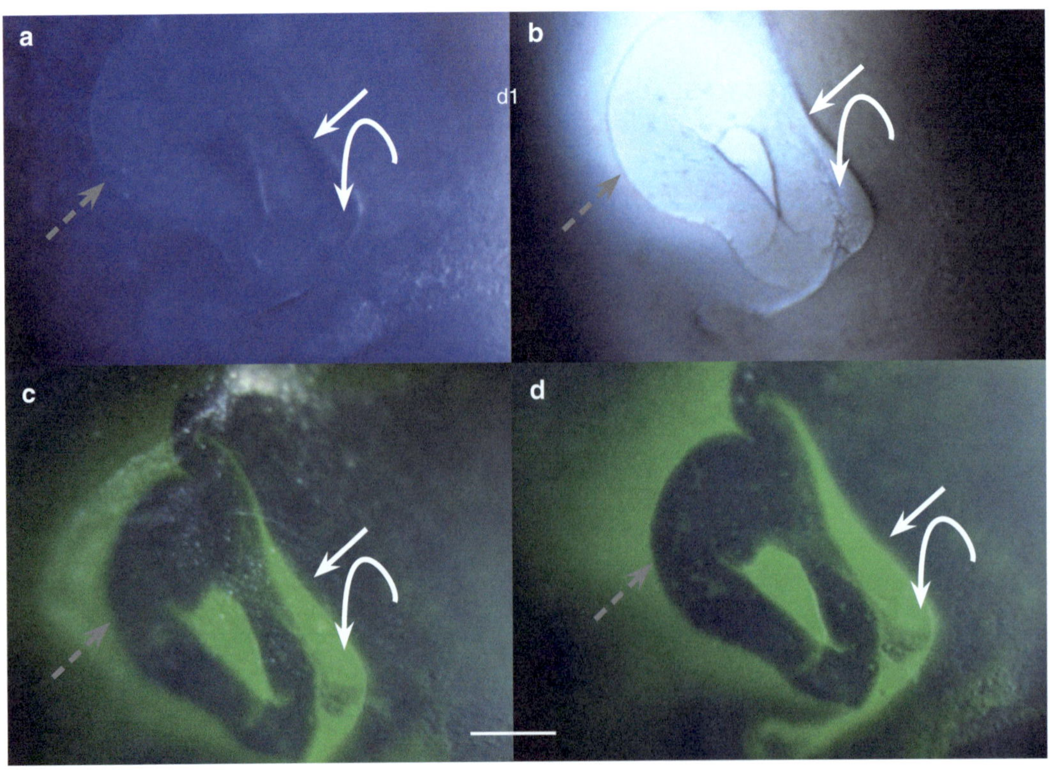

Fig. 2.38 Survey of a peculiar epithelial 'flap' (*dashed arrows*), (**a**) greyish in focal and (**b**) translucent in strong retroillumination, that (**c, d**) does not stain with fluorescein. The *plain arrows* indicate an epithelial edge to the right from the 'flap' and the *bowed arrows* a space between them in which (**c, d**) fluorescein is pooling (the photographs are shown in the order as taken; the arrows are placed in corresponding locations)

Case 12. Surface Erosions in Predisposed Cornea (2) (cont.)

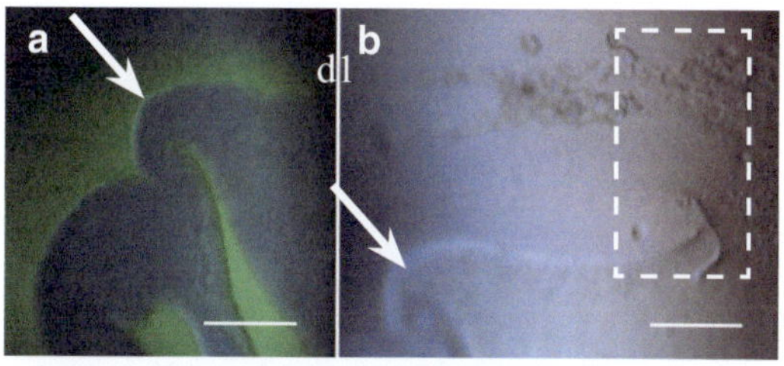

Fig. 2.39 Survey of the upper part of the 'flap' shown in Fig. 2.38. The *arrows* indicate the same location. The area in *frame* is shown in Fig. 2.40b (below) (the arrows are placed in corresponding locations)

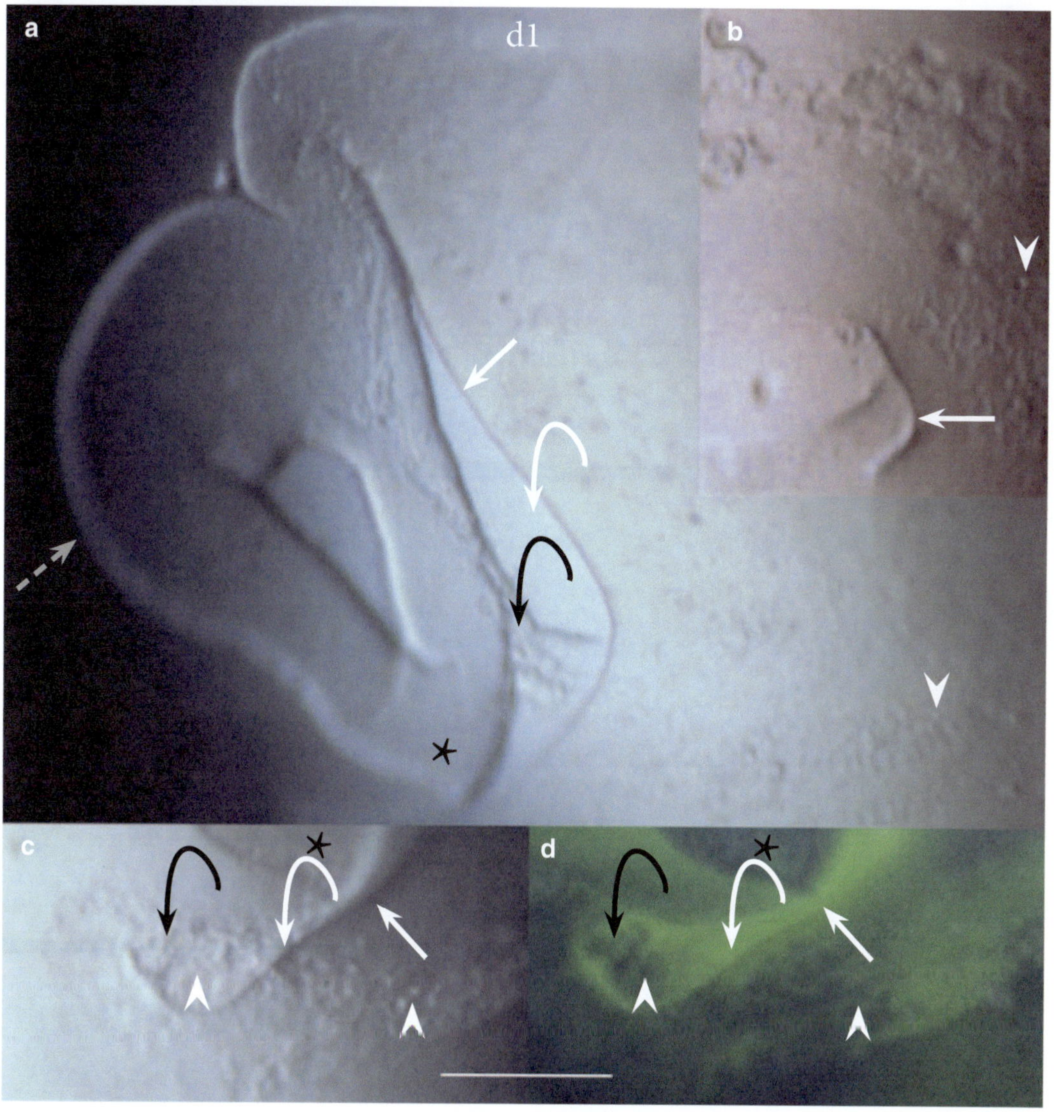

Fig. 2.40 (a) The space (*white bowed arrow*) between the 'flap' (*dashed arrow*) and the epithelial edge to the right (*plain arrow*) appears smooth except for some tissue that is slightly elevated (*black bowed arrow*; cf. Fig. 2.38c, d in which it appears dark). In the surroundings in places rounded bodies (*arrowhead*) are visible; the *asterisk* indicates the same location as in (c, d). (b) shows epithelial derangement in the adjacent area comprising rounded bodies (*arrowhead*) and epithelial lines (*arrow*). (c, d) show the space (*white bowed arrows*) adjacent to the lower tip of the 'flap' (*asterisks*); as in (a), the space contains some tissue (*black bowed arrows*) that appears elevated (dark) in the green stained fluid; the *arrowheads* indicate rounded bodies (the markers in (c, d) are placed in corresponding locations)

Case 12. Surface Erosions in Predisposed Cornea (2) (cont.)

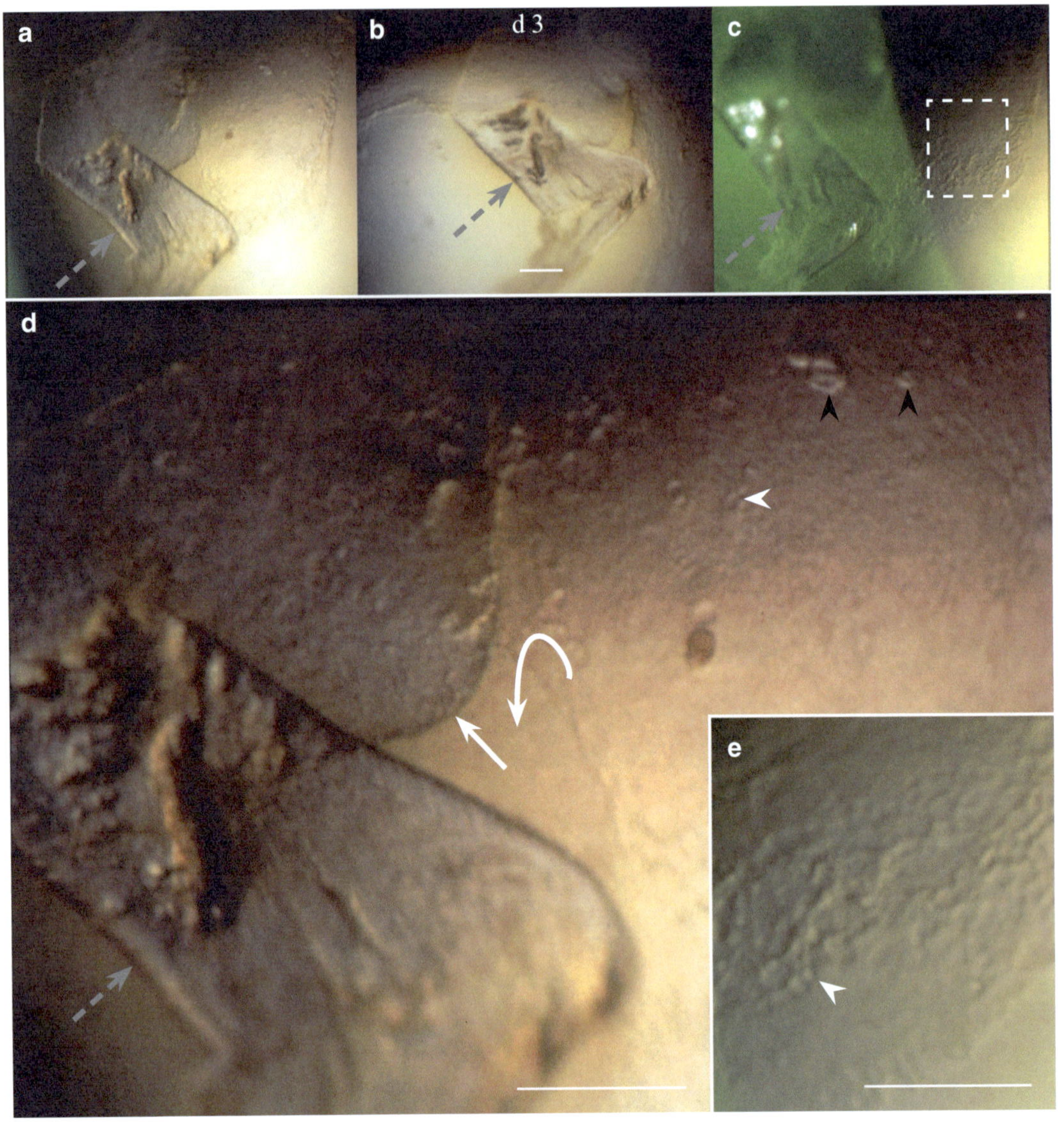

Fig. 2.41 (a–c) Survey of a loose epithelial flap (*arrow*) moving on the surface; the area indicated by *frame* in (c) is shown at higher magnification in (e). (d) The bottom of the erosion (*bowed arrow*) close to the flap (*dashed arrow*) appears smooth and the surrounding epithelium (*plain arrow*) shows many rounded bodies (*white arrowhead*) and some cysts (*black arrowheads*). (e) The area in *frame* in (c) shows many rounded bodies (*white arrowhead*)

Comment
The rounded bodies in Figs. 2.40 and 2.41 probably represent swollen epithelial cells.

Case 12. Surface Erosions in Predisposed Cornea (2) (cont.)

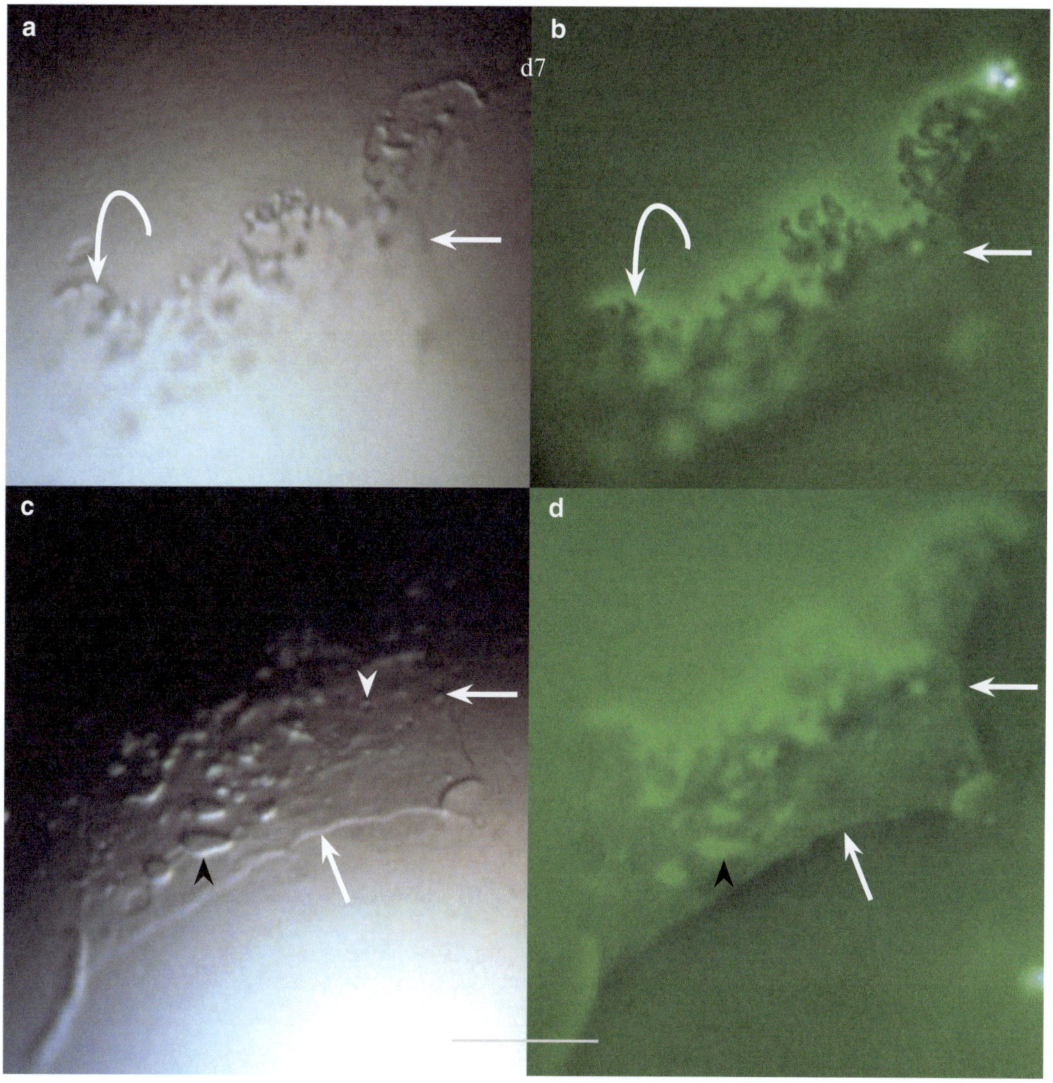

Fig. 2.42 (**a**, **b**) The upper edge of the remaining lesion is irregular and (**b**) fluorescein is pooling in front of it (*bowed arrowheads*) (cf. Fig. 1.22); the *straight arrowheads* indicate a smooth curved edge to the right. (**c**, **d**) The lower edge of the same lesion appears almost smooth. The lesion contains rounded bodies (*white arrowhead*) and cysts (*black arrowheads*) (the markers are placed in corresponding locations in each pair of photographs and the right straight arrows in all of them)

Case 12. Surface Erosions in Predisposed Cornea (2) (cont.)

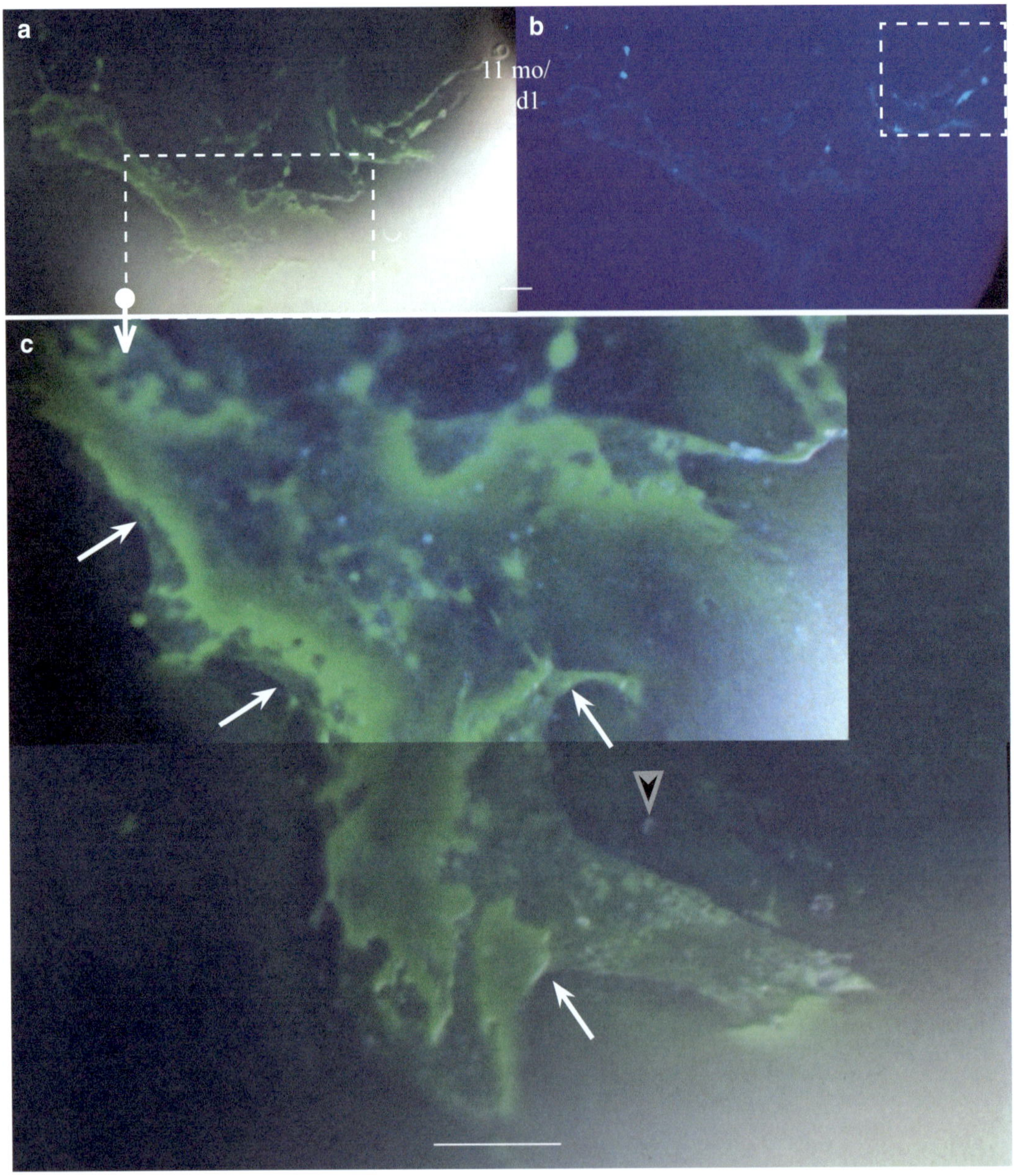

Fig. 2.43 (a, b) Survey of a spontaneously deranged epithelial area visualised with (a) fluorescein and (b) fluorescein and blue filter; the area in *frame* in (b) is shown at higher magnification in Fig. 2.44. (c) The area in *frame* in (a) shows curved edges (*arrows*) and various shades of green staining. In the surroundings a few cysts (*arrowhead*) are visible. For some details see also Fig. 1.28 (c is a composed photograph)

Case 12. Surface Erosions in Predisposed Cornea (2) (cont.)

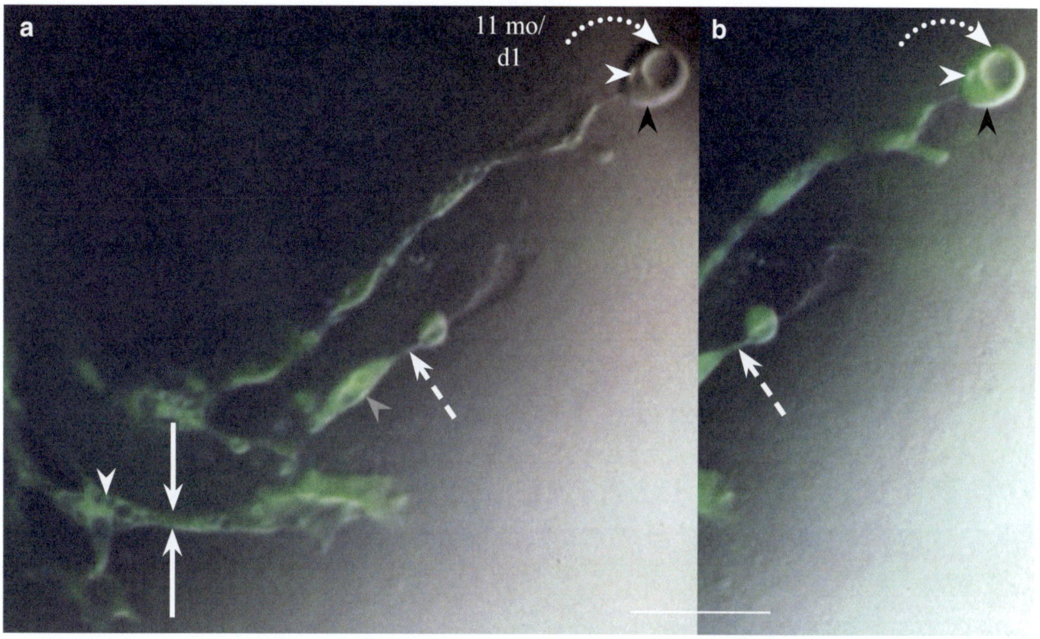

Fig. 2.44 (a) In the *right upper part* of the area shown in Fig. 2.43b edges of epithelial sheets (*plain arrows*) are visible and between them are clefts containing rounded bodies (left white arrowhead), fusiform expansions (grey arrowhead) and epithelial lines (dashed arrow). In the right upper corner is visible a rounded cystic space (*black arrowhead*) that contains a rounded body (*white arrowhead*) and a larger rounded structure (*bowed arrow*; see also Figs. 1.3h and 1.39). (b) shows green fluorescein staining of the *right upper part* of the area (the markers are placed in corresponding locations)

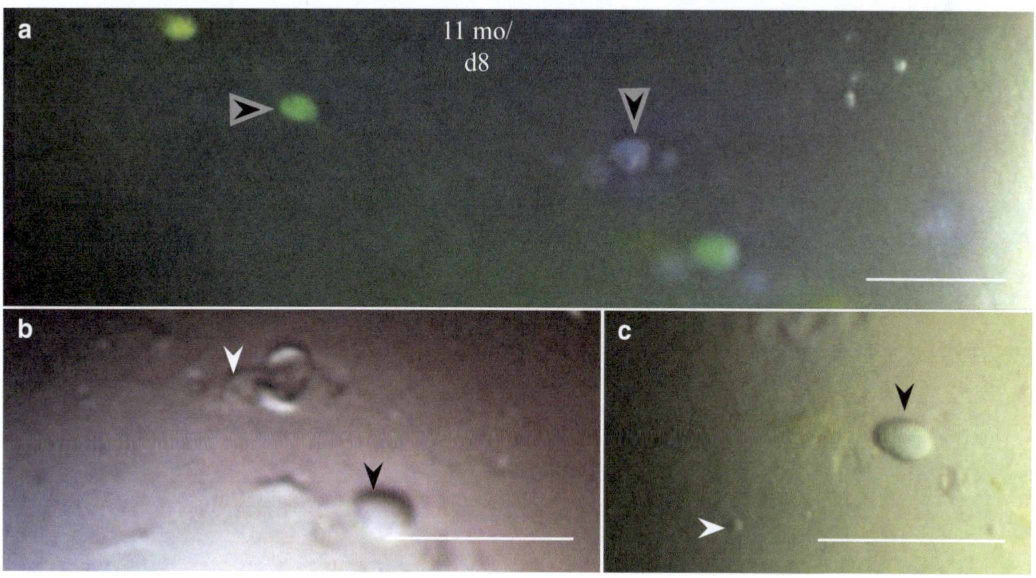

Fig. 2.45 A week later, the deranged area shows cysts (*black arrowheads*) and rounded bodies (*white arrowheads*). In (a), (b, c) with fluorescein, it is visible that some but not all cysts stain green

Case 13. The Dynamics of Cysts and Maps in EBMD

Case Report

A 33-year-old man presented because of slight grittiness in his right eye hit by a small tree branch on the previous day. Both corneae showed pronounced greyish map-like changes and the upper temporal quadrant of the right cornea a small erosion, which healed rapidly. He recalled a similar event about 4 years previously. During the following almost 7 years, he noted fluctuating vision caused by a slight irregular astigmatism but no symptoms suggestive of recurrent erosions. The patient's father had similar corneal epithelial changes (but never any symptoms). Of his two younger sisters one had occasionally slight irritative symptoms, and her corneae showed various numbers of epithelial cysts but no other detectable changes; the other sister had no symptoms and her corneae appeared normal.

The photographs of the right cornea were taken on day 1 (at presentation) and at several occasions during the following 7 years.

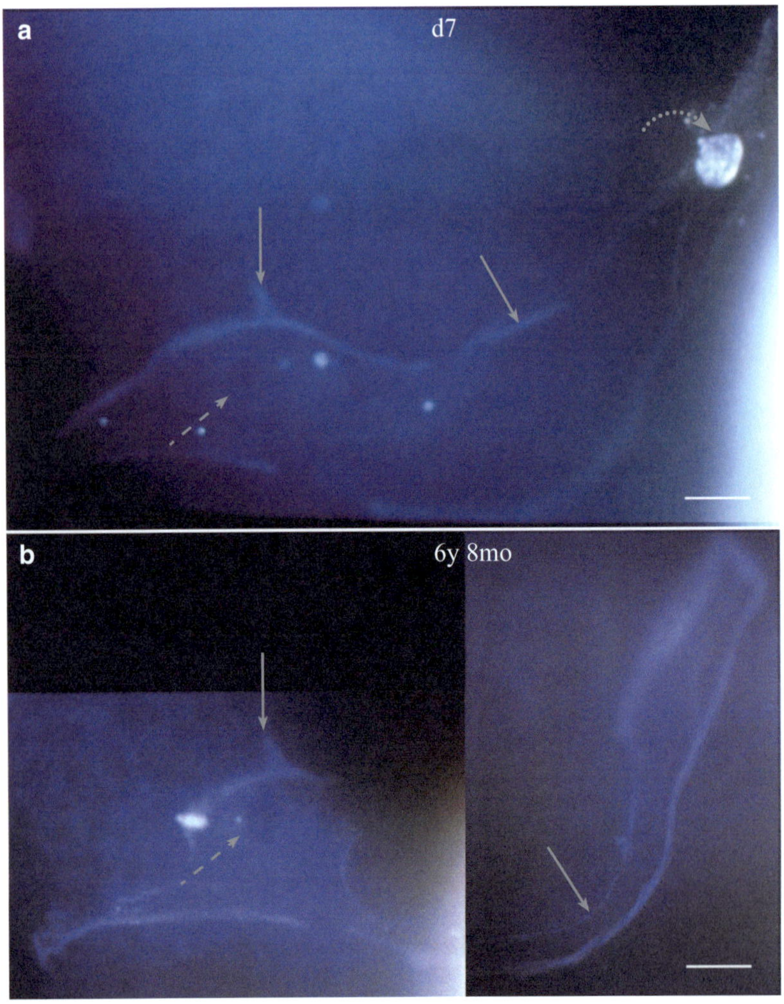

Fig. 2.46 Survey of a greyish membrane ('map') localised in the lower nasal cornea. It is delineated by brighter lines (*plain arrows*) and contains darker 'holes' (*dashed arrows*). (**a**) additionally shows a strongly light-reflecting cyst (*bowed arrow*, cf. Fig. 2.47). In (**b**) it is shown that with time the upper line (*plain arrows*) has approached the lower one (for details see Figs. 2.50 and 2.51; (**b**) is a composed photograph)

Case 13. The Dynamics of Cysts and Maps in EBMD (cont.)

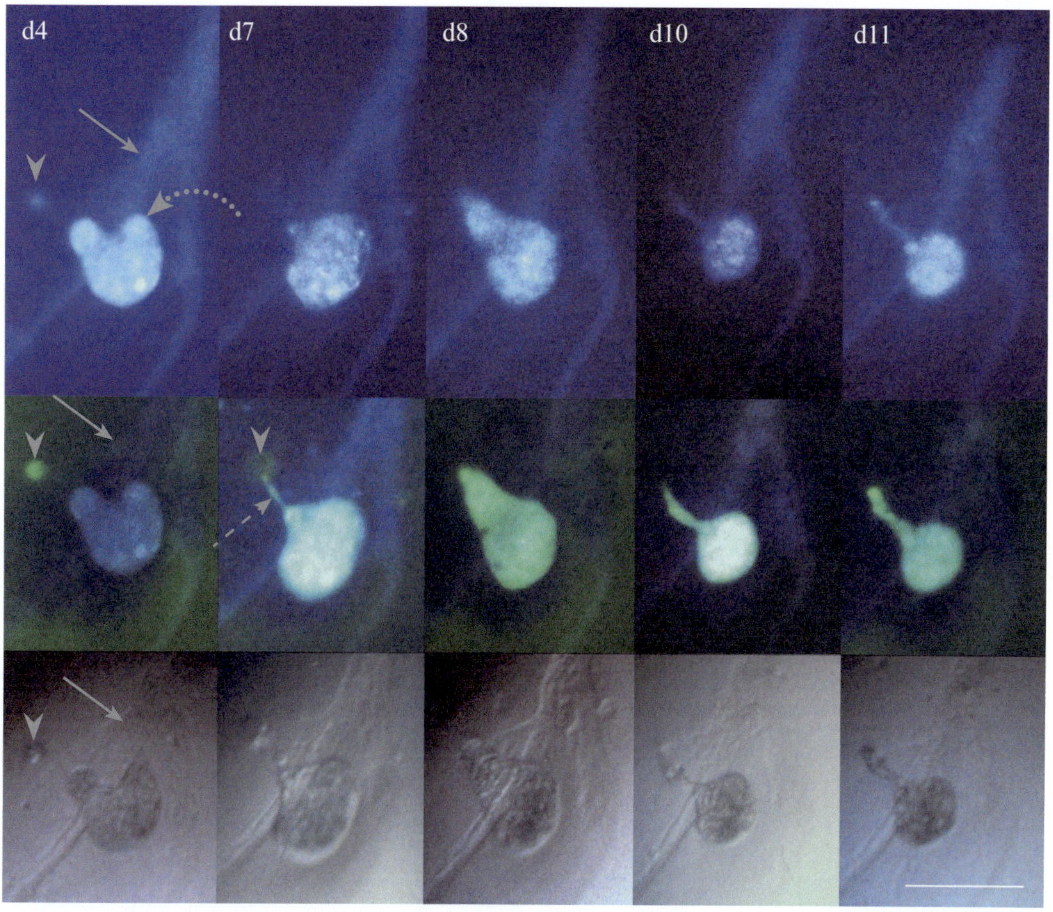

Fig. 2.47 Survey of development of the large cyst (*bowed arrow*), shown in Fig. 2.46a, between days 4 and 11; the *arrowheads* indicate a small cyst in apposition and the *plain arrows* a light-reflecting edge of a 'map'. The *first row* shows the strong light-reflecting property of the large cyst. In the *second row* it is visible that it does not stain green with fluorescein on day 4 but does so on days 7–11; the green staining visualises that a small channel (d7, *arrow*) connecting the large cyst with the location of the small one (d7, *arrowhead*) has developed between d4 and d7. The channel enlarges (d8) and becomes smaller again (d10–d11), and the cyst diminishes in size. The *third row* shows in retroillumination how the cyst forces its way toward the surface through or below the line (*arrow*) (for additional details see Figs. 2.48 and 2.49, overleaf)

Case 13. The Dynamics of Cysts and Maps in EBMD (cont.)

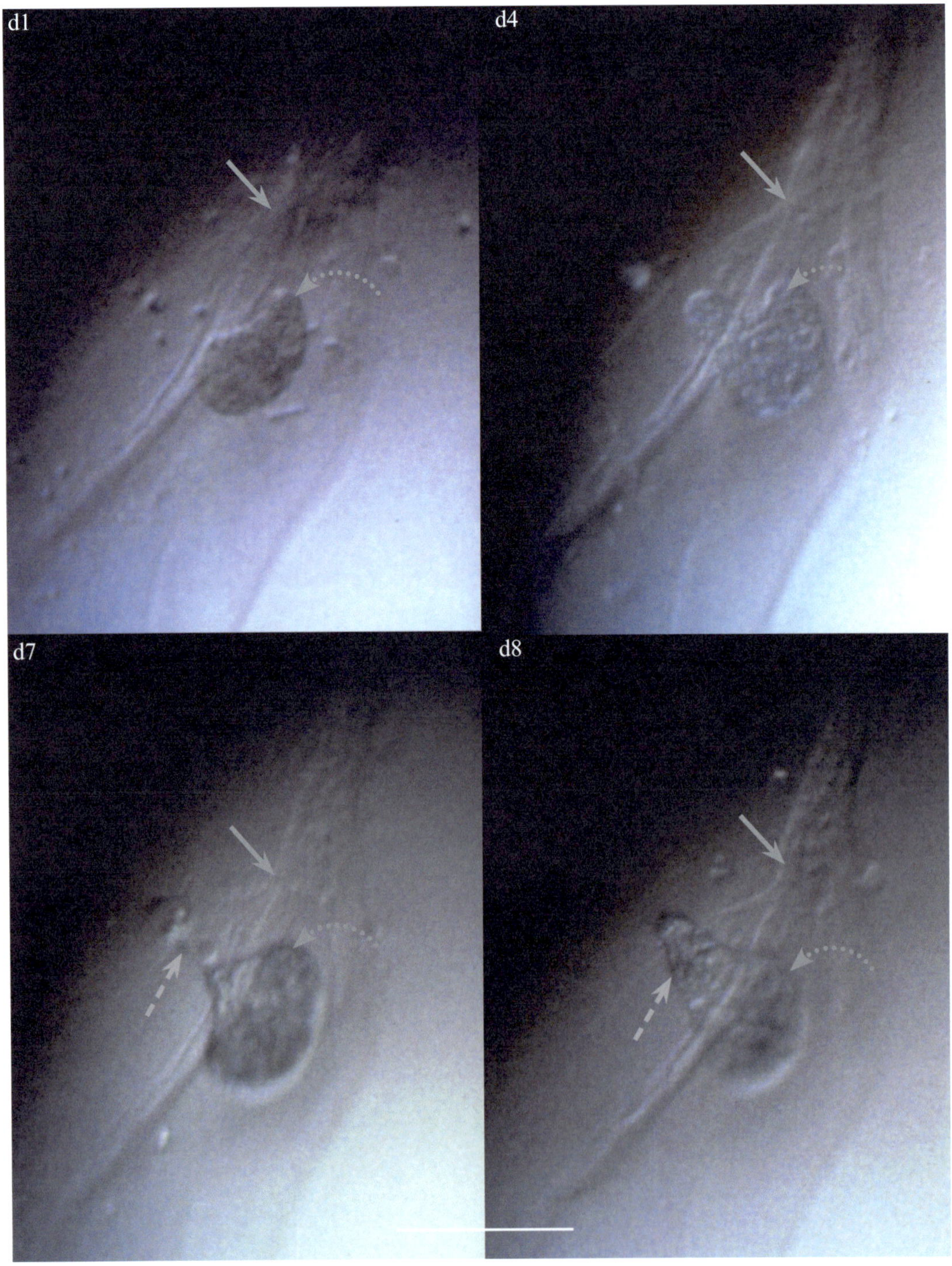

Fig. 2.48 Dynamics of the cyst (*bowed arrows*), shown in Figs. 2.46 and 2.47, between days 1 and 8. It forces its way toward the surface below or between different layers of the edge (*plain arrows*) of the 'map' shown in Fig. 2.46a. The contents of the cyst appear granular. On d7 a channel (*dashed arrow*) connecting the cysts with the surface is visible (cf. Fig. 2.47) (the plain arrows are placed in approximately corresponding locations)

Case 13. The Dynamics of Cysts and Maps in EBMD (cont.)

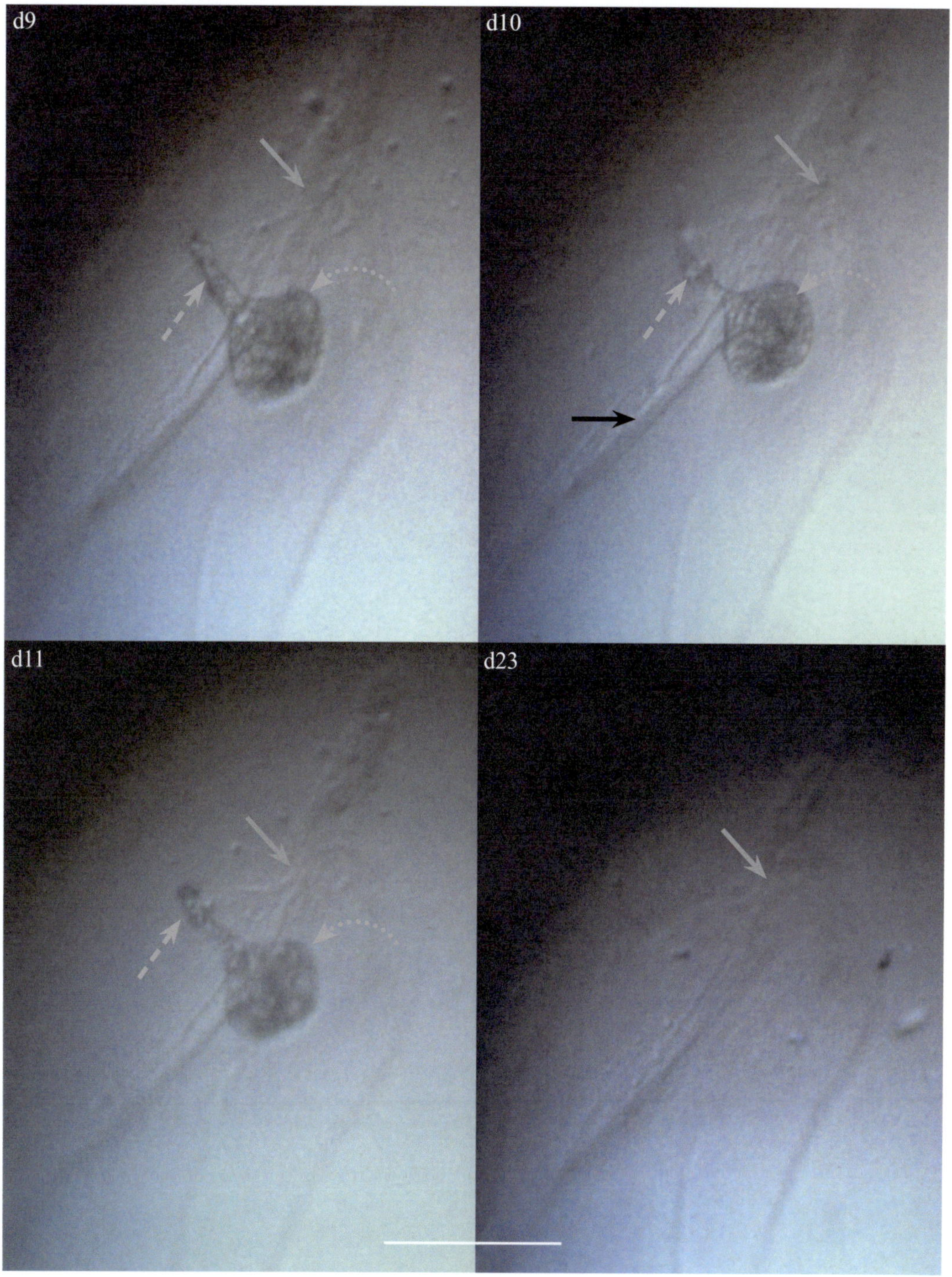

Fig. 2.49 Further development of the cyst (*bowed arrows*) shown in Fig. 2.48, captured between days 9 and 23. It seems to slowly empty its contents through the channel (*dashed arrows*). The cyst has disappeared between d11 and d23 (as in Fig. 2.48, the plain arrows indicating the edge of the 'map' shown in Fig. 2.46 are placed in approximately corresponding locations)

Case 13. The Dynamics of Cysts and Maps in EBMD (cont.)

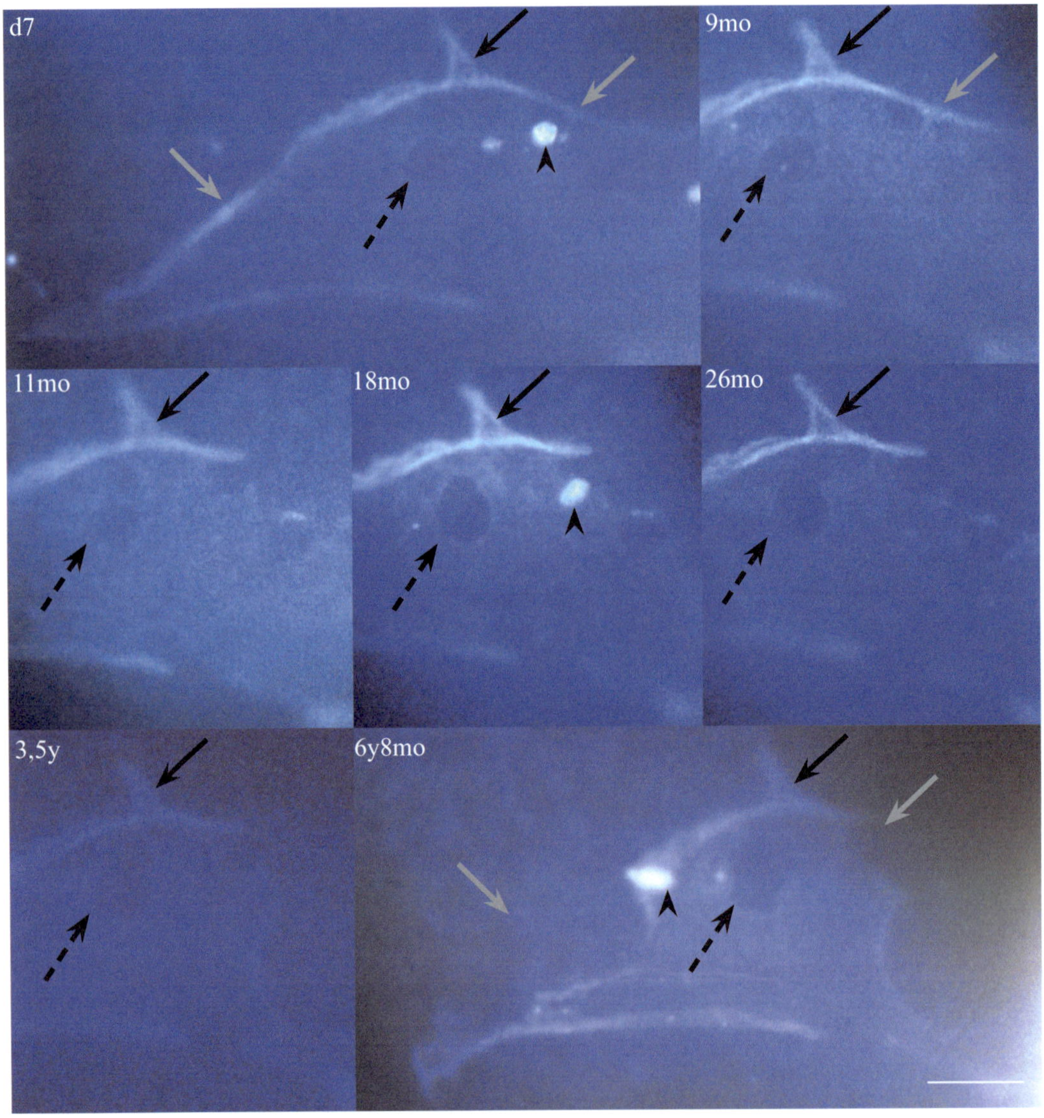

Fig. 2.50 Changes in a greyish 'map' and its edges (*plain black arrows*, cf. Fig. 2.46) captured at various occasions between day 7 and after <7 years. The dark 'hole' (*dashed arrows*) within the 'map' does not change much. The upper edge of the 'map' (*grey arrows*) moves in places toward the lower one. Some light-reflecting cysts (*arrowheads*) are coming and going (the arrows are placed in corresponding locations)

Case 13. The Dynamics of Cysts and Maps in EBMD (cont.)

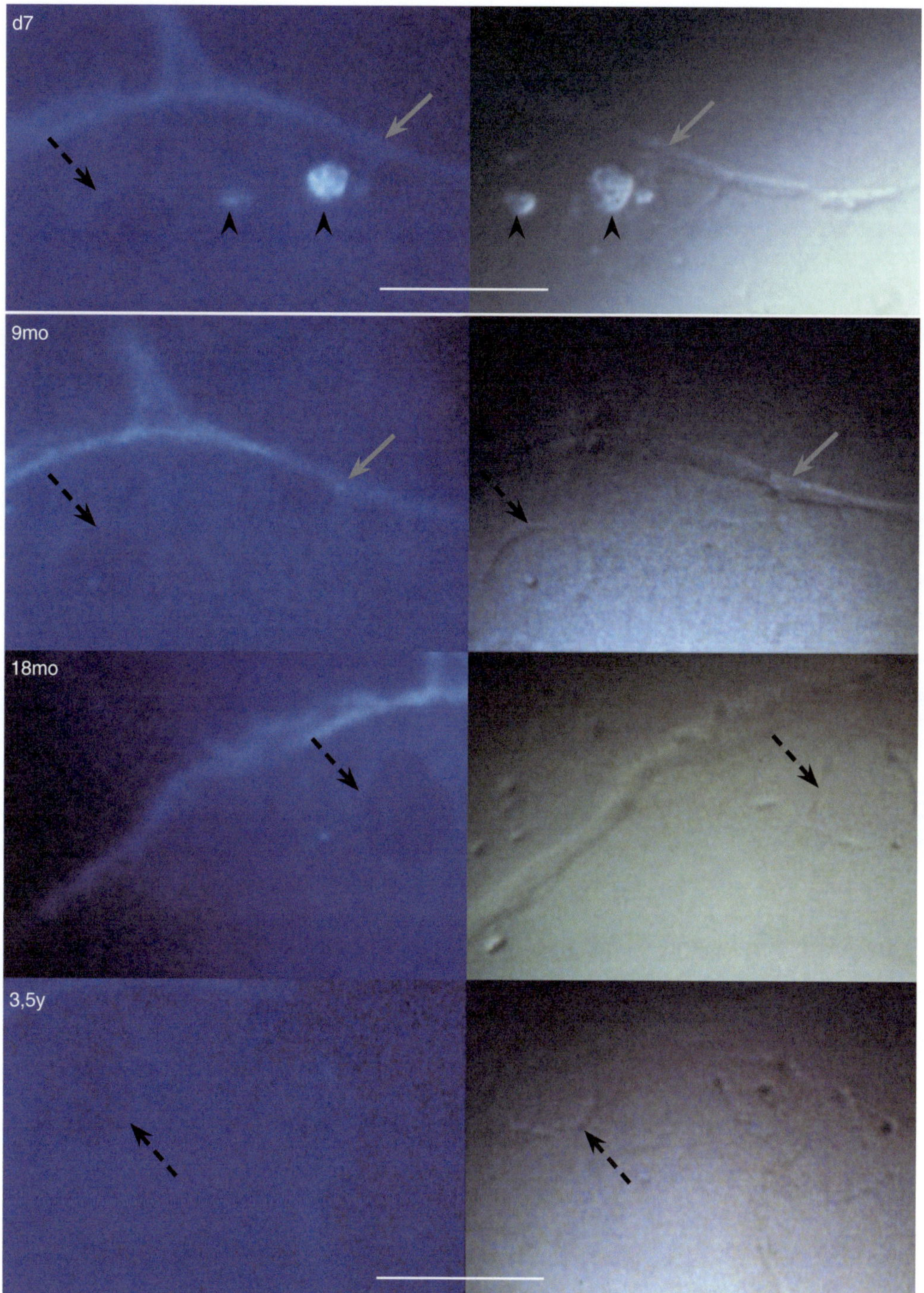

Fig. 2.51 The *left upper part* of the 'map' shown in Fig. 2.46. On d7, it shows light-reflecting cysts (*arrowheads*); the *grey arrows* indicate a light-reflecting (*left column*), thickened (*right column*) edge of the 'map' and the *dashed arrow* a 'hole' within it. Photographs taken at 9 and 18 months and 3.5 years show that the 'hole' (left column, *dashed arrows*) has a thickened edge (right column, *dashed arrows*) (the markers are placed in corresponding locations in each pair of photographs)

Case 13. The Dynamics of Cysts and Maps in EBMD (cont.)

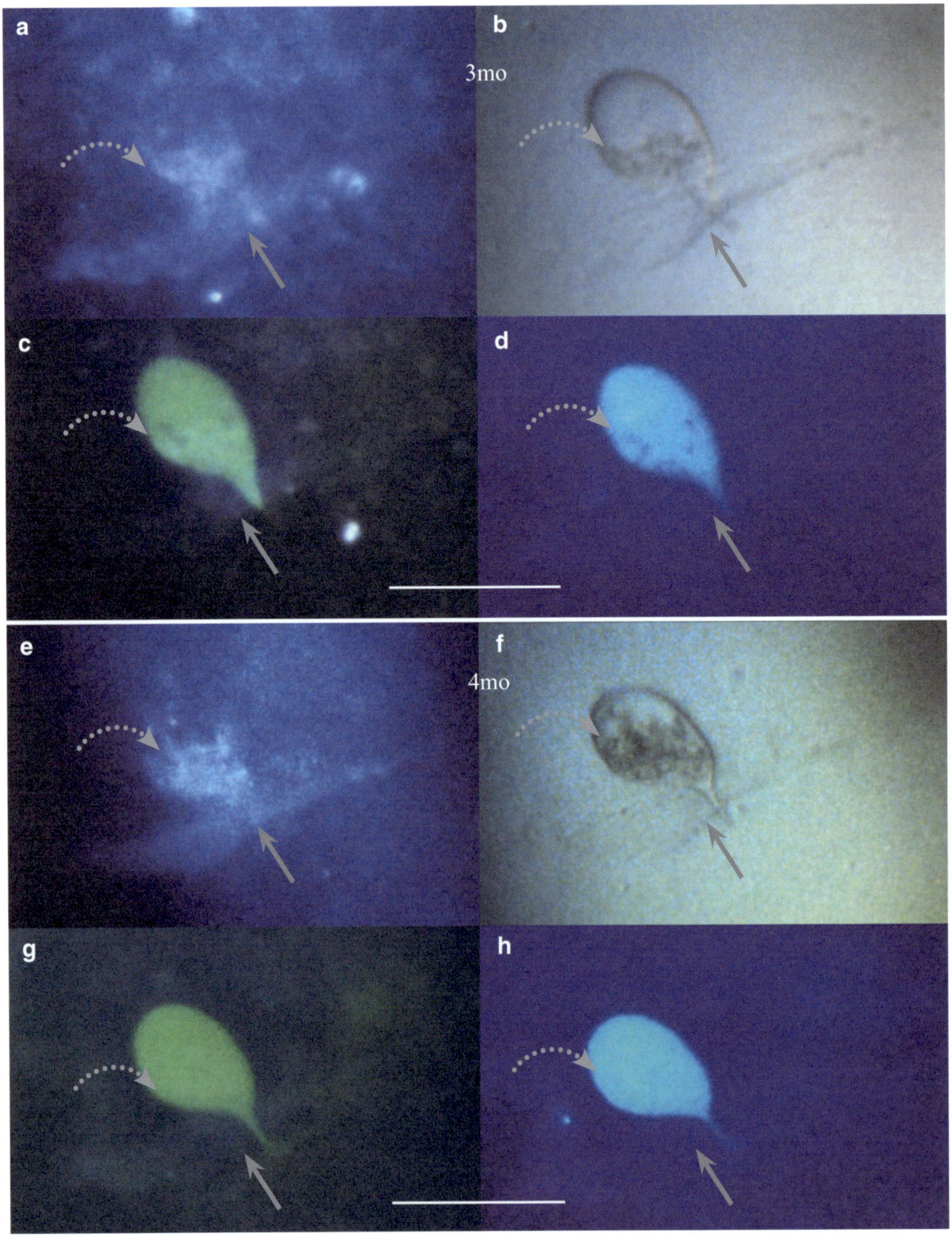

Fig. 2.52 (a–d) A cyst (*bowed arrows*), adjacent to a thickened linear structure (*plain arrows*), captured in the lower temporal part of the cornea at 3 months. It shows (a) partly light-reflecting contents which (b) fill its lower part; the cyst seems to force its way toward the surface below or through the linear structure. (c, d) The green fluorescein staining indicates a connection with the surface. (e–h), captured one month later, show a cyst in the same location and with the same properties. See also Fig. 1.15 (the markers are placed in corresponding locations in each set of photographs; (d, h) blue filter)

Case 13. The Dynamics of Cysts and Maps in EBMD (cont.)

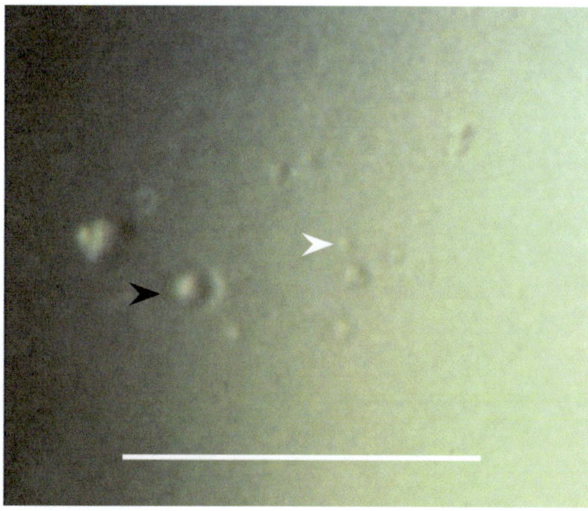

Fig. 2.53 Rounded cysts (*black arrowhead*) and rounded bodies (*white arrowhead*) captured at one occasion

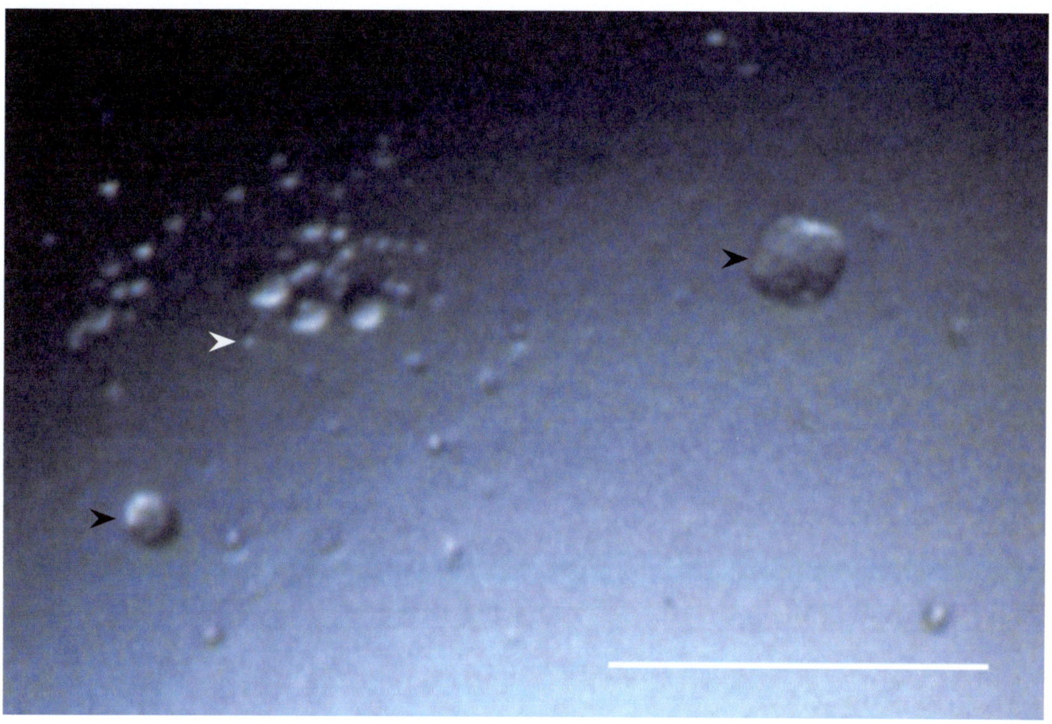

Fig. 2.54 Larger and smaller cysts (*black arrowheads*) and rounded bodies (*white arrowhead*) captured in one of the patient's sisters

3 Recurrent Erosions and Stromal Infiltrates

In recurrent erosions, stromal infiltrates are rare. The 3 patients presented in this chapter had previously known recurrent erosions in the involved eye, and 2 of them (Cases 1 and 3) had epithelial changes indicating propensity to recurrent erosions in the fellow eye. One cornea (Case 1) showed an infiltrate at presentation. In the remaining 2, infiltrates developed after onset of a new, spontaneous erosion; in both, they developed shortly after topical antibiotic treatment was started. Microbiological work-up did not reveal any relevant bacteria and the infiltrates resolved with addition of topical steroid. The infiltrates left superficial scars that, out of the visual axis, did not impair the visual acuity. In 1 patient (Case 3), shortly after the steroid was stopped, new infiltrates developed close to the scar.

The follow-up series performed in 2 patients (Cases 2 and 3), starts after 3 days of topical cortisone treatment. At that stage, the infiltrates were superficial and appeared either as a rounded whitish fleck measuring about 500 μm in diameter or lobulated one (with largest diameter of about 1.2 mm). The infiltrates contained large numbers of rounded bodies (here interpreted as invading inflammatory cells) and showed rounded central foci. In both patients, the resolution followed the same path; the central foci were first to clear. Some rounded bodies were still present within the areas of scars 6 weeks (Case 2) and 2 months (Case 3) after onset.

H.M. Tabery, *Recurrent Erosion Syndrome and Epithelial Edema: In Vivo Morphology in the Human Cornea*,
DOI 10.1007/978-3-319-06545-8_3,

Case 1. Stromal Infiltrates 2.5 Years After Onset of Recurrent Erosions

Case Report

A 46-year-old woman with recurrent erosion symptoms in the left eye after a minor injury about 4 months previously. At presentation, the paracentral lower cornea showed an area of deranged epithelium. The right cornea appeared normal. She was treated with NaCl ointment 5 % at bedtime. A month later, the paracentral lower cornea still showed epithelial derangement, but after a further month only a few epithelial cysts were present. After that, symptoms and findings were minimal; at one occasion, small cysts were noted in both paracentral lower corneae. Ten months after the last visit (about 2, 5 years after the onset of symptoms), she presented 4 days after onset of new symptoms with an infiltrate in the left paracentral lower cornea, accompanied by epithelial edema and anterior uveitis. Conjunctival smears showed no bacteria. Initial treatment with topical antibiotics had no effect, but the symptoms rapidly ameliorated with topical dexamethasone 0.1 % twice a day. The infiltrate resulted in a superficial stromal scar.

The patient additionally suffered from recurrent bouts of episcleritis in the left eye and carpal tunnel syndrome.

The photographs of the left cornea were taken on day 1 (at presentation) and 1, 2, 10 and 19 months later and of the infiltrate 10 days after cortisone treatment was started.

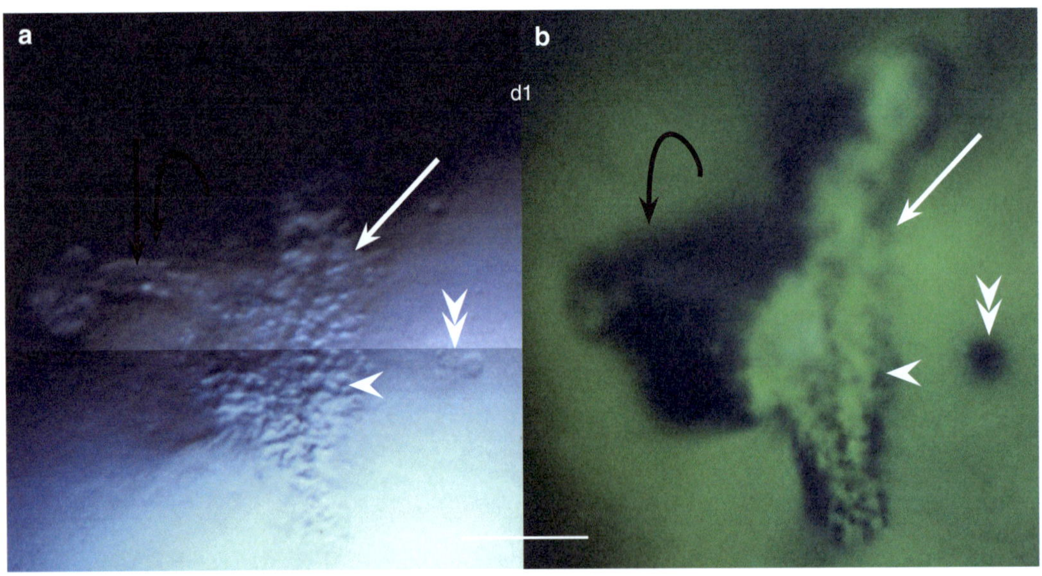

Fig. 3.1 (**a**) A larger lesion (*arrow*) containing many rounded bodies (*arrowhead*) and a small adjacent one (*double arrowhead*). (**b**) Some parts of the lesion (*bowed arrow*), the adjacent one (*double arrowhead*) and some or the rounded bodies (*arrowhead*) are protruding (dark) in the green stained tear film (The arrowheads and the straight arrows are placed in corresponding locations)

Case 1. Stromal Infiltrates 2.5 Years After Onset of Recurrent Erosions (cont.)

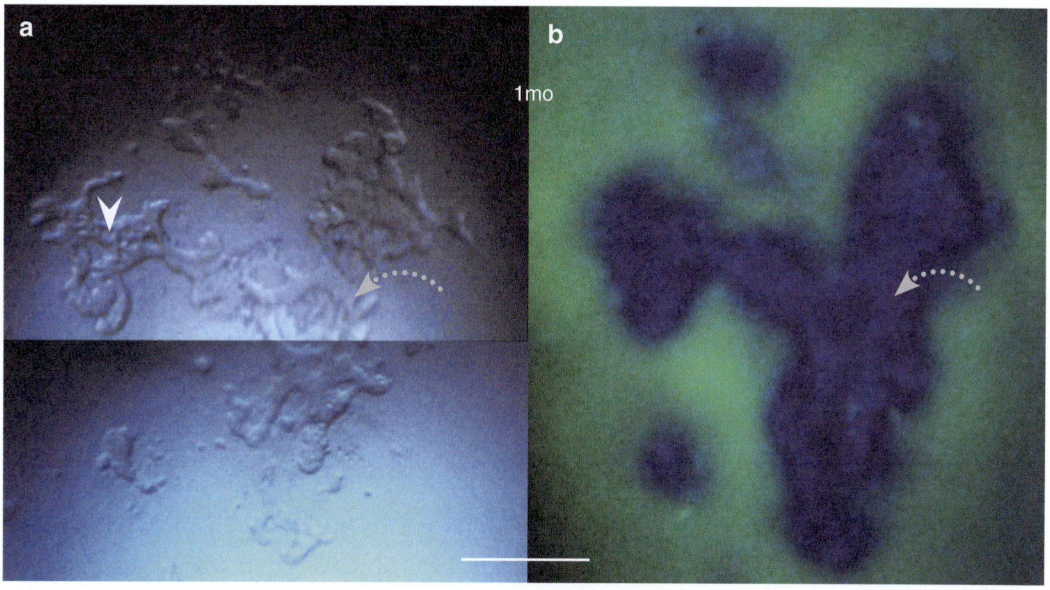

Fig. 3.2 (a) Smaller and larger lesions in apposition; visible are rounded bodies (*arrowhead*) and confluent structures (*bowed arrow*) with no discernible details. (b) The abnormal structures (*bowed arrow*) are light-reflecting (greyish) and protruding (dark) in the tear film stained green with fluorescein. (The arrows are placed in corresponding locations. (a) is a composed photograph.) (see also Figs. 1.1 and 2.17)

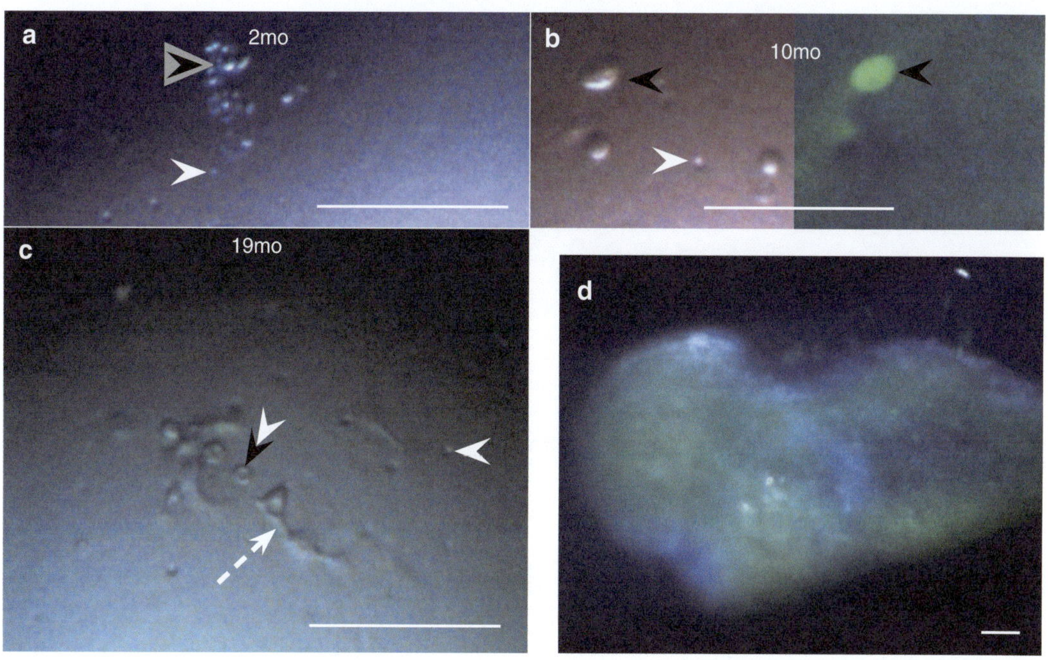

Fig. 3.3 (a–c) At various occasions, the epithelium showed rounded bodies (*white arrowheads*), cysts (*black arrowheads*), some (b) staining green with fluorescein and some (c) containing rounded bodies (*double arrowhead*); (c) additionally shows epithelial lines (*arrow*; cf. Fig. 1.24). (d) Infiltrate captured after 10 days of cortisone treatment; its lobulated shape implies confluence of 2, possibly 3 rounded infiltrates in apposition (cf. Cases 2 and 3)

Case 2. Stromal Infiltrates During the First Episode of Recurrent Erosion

Case Report

A 40-year-old man presented with pain in the left eye struck by a leaf 2 days previously. The paracentral lower cornea showed an epithelial erosion that healed with antibiotic ointment. Three months later, the cornea suffered a new, spontaneous erosion in the same location. The right cornea appeared normal. Within 2 days of treatment with chloramphenicol ointment developed two subepithelial infiltrates in the paracentral lower cornea. After a further 2 days with neomycin-bacitracin ointment, the infiltrates appeared unchanged, but they started to resolve with addition of prednisolone acetate drops 0.5 % twice a day. Conjunctival smears revealed *Staphylococcus aureus* sensitive to the used antibiotics.

The patient had chronic blepharitis, bouts of urticaria and a family history of allergy.

The photographic series starts on day 1 (after 3 days of cortisone treatment) and follows the resolution of the infiltrates captured on days 3, 8 and 15 and after 6 weeks.

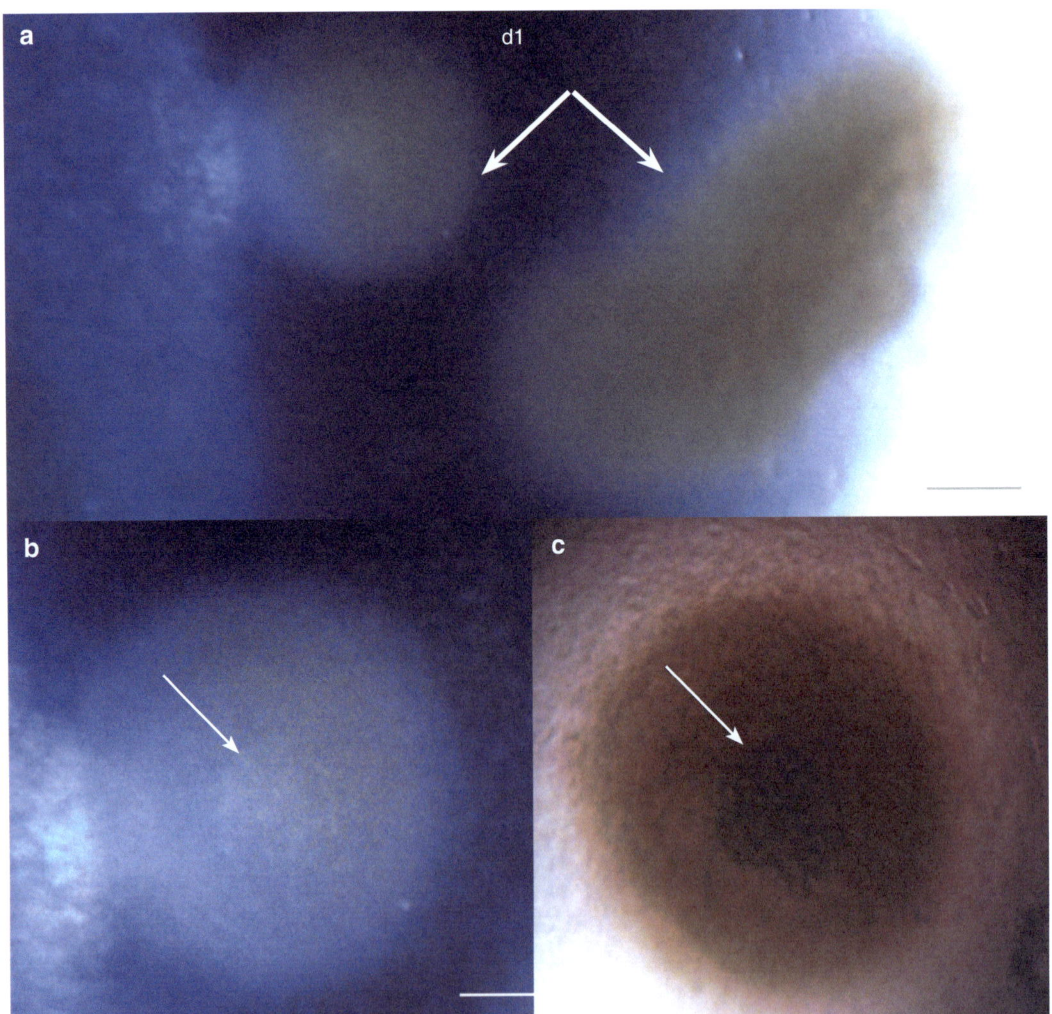

Fig. 3.4 (**a**) Two infiltrates (*arrows*) in apposition. (**b**) shows the light-reflecting property of the left infiltrate; the *arrow* indicates its rounded centre which in (**c**, *arrow*) appears darker than its surroundings. (The arrows in **b** and **c** are placed in corresponding locations)

Case 2. Stromal Infiltrates During the First Episode of Recurrent Erosion (cont.)

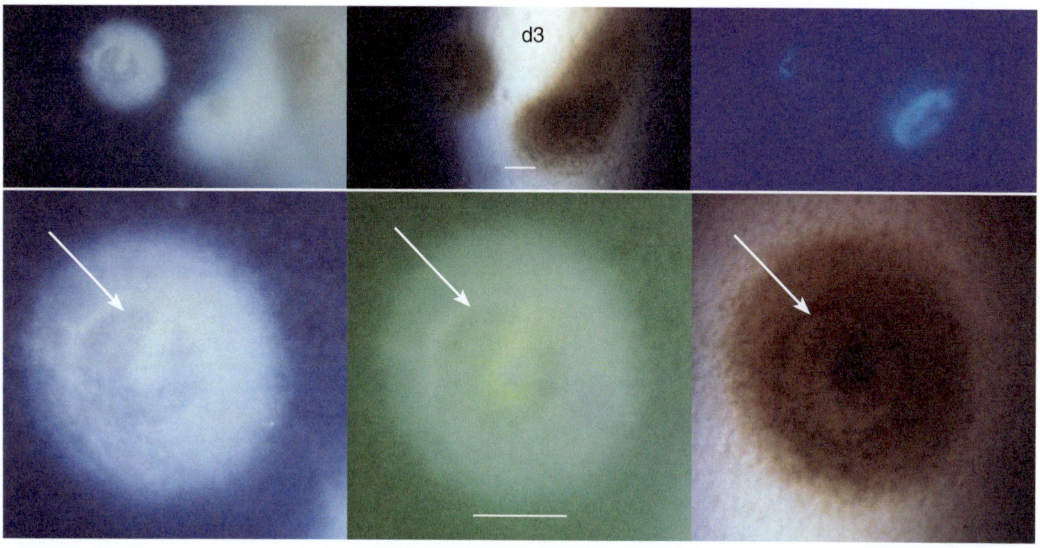

Fig. 3.5 *Upper row*: Survey; the image to the right shows fluorescein staining in the centre of the left infiltrate and in the lower part of the right one (blue filter). *Lower row*: The left infiltrate starts to clear in the rounded central area (*arrows*, placed in corresponding locations)

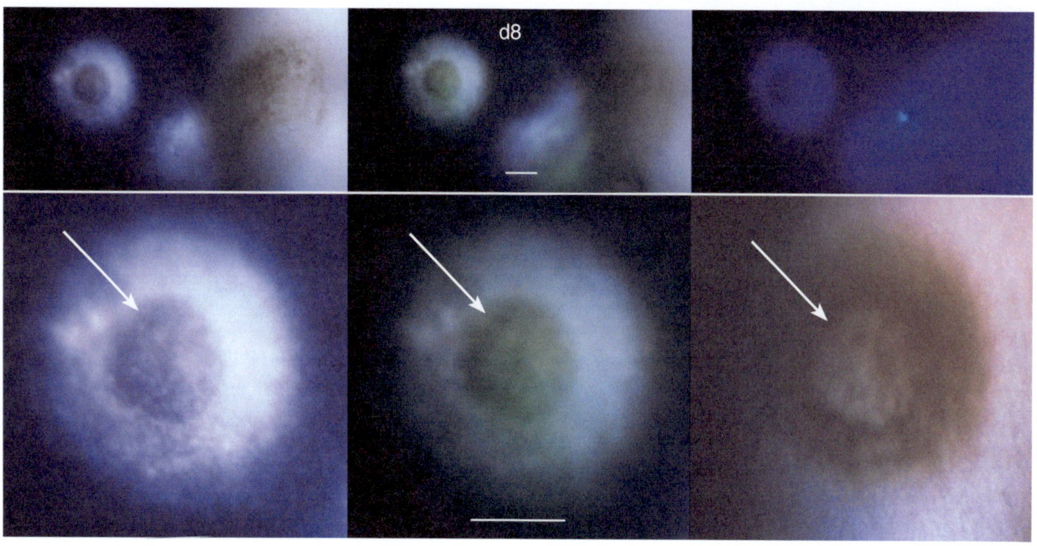

Fig. 3.6 *Upper row*: Survey; the image to the right shows no fluorescein staining of the left infiltrate; a small area stains in the right one (blue filter). *Lower row*: ongoing clearing of the rounded central area (*arrows*, placed in corresponding locations)

Case 2. Stromal Infiltrates During the First Episode of Recurrent Erosion (cont.)

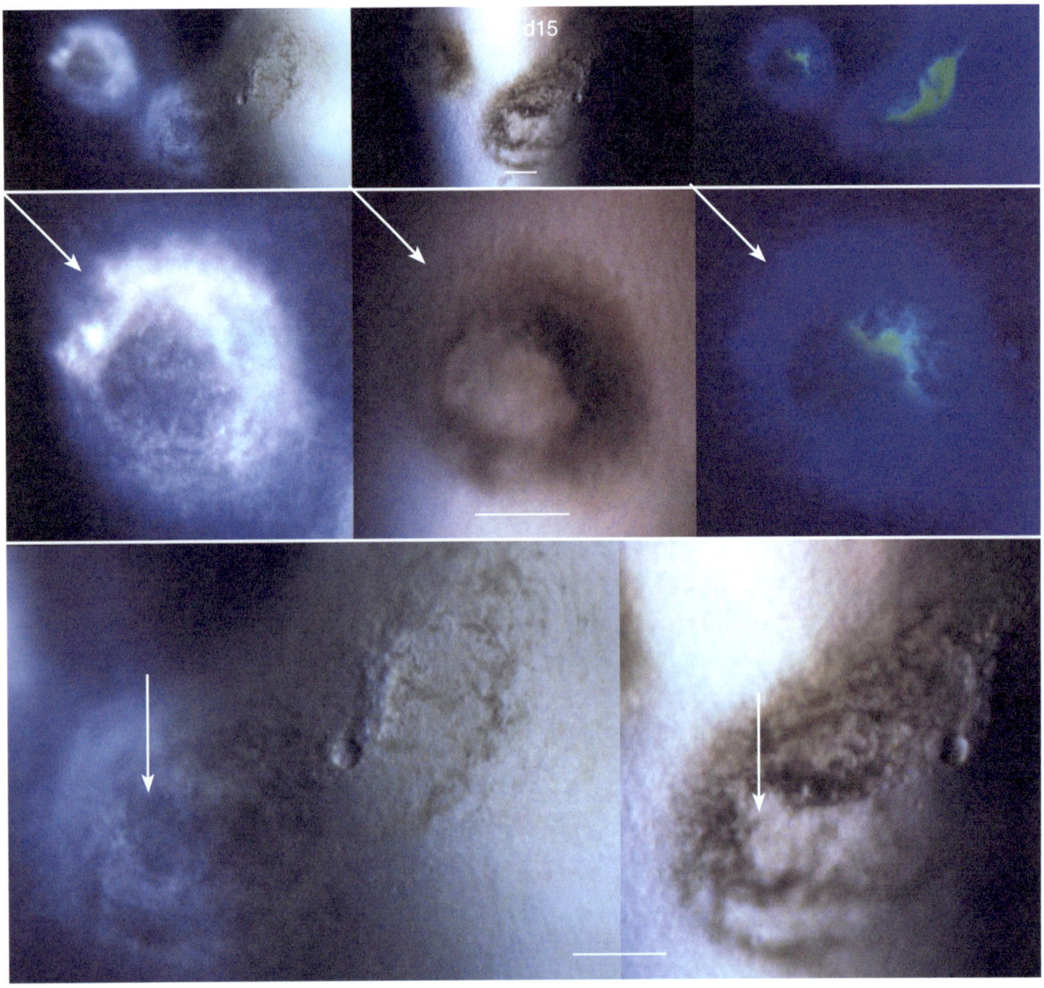

Fig. 3.7 *Upper row*: Survey. The image to the right shows fluorescein staining in the centre of the left infiltrate and in the lower and central part of the right one (blue filter). The *central row* shows ongoing clearing of the left infiltrate (*arrows*) and to the right fluorescein pooling in its central area (blue filter). In the *lower row* is visible clearing (*arrows*) of the left lower part of the right infiltrate. (The arrows are placed in corresponding locations in each set of photographs)

Case 2. Stromal Infiltrates During the First Episode of Recurrent Erosion (cont.)

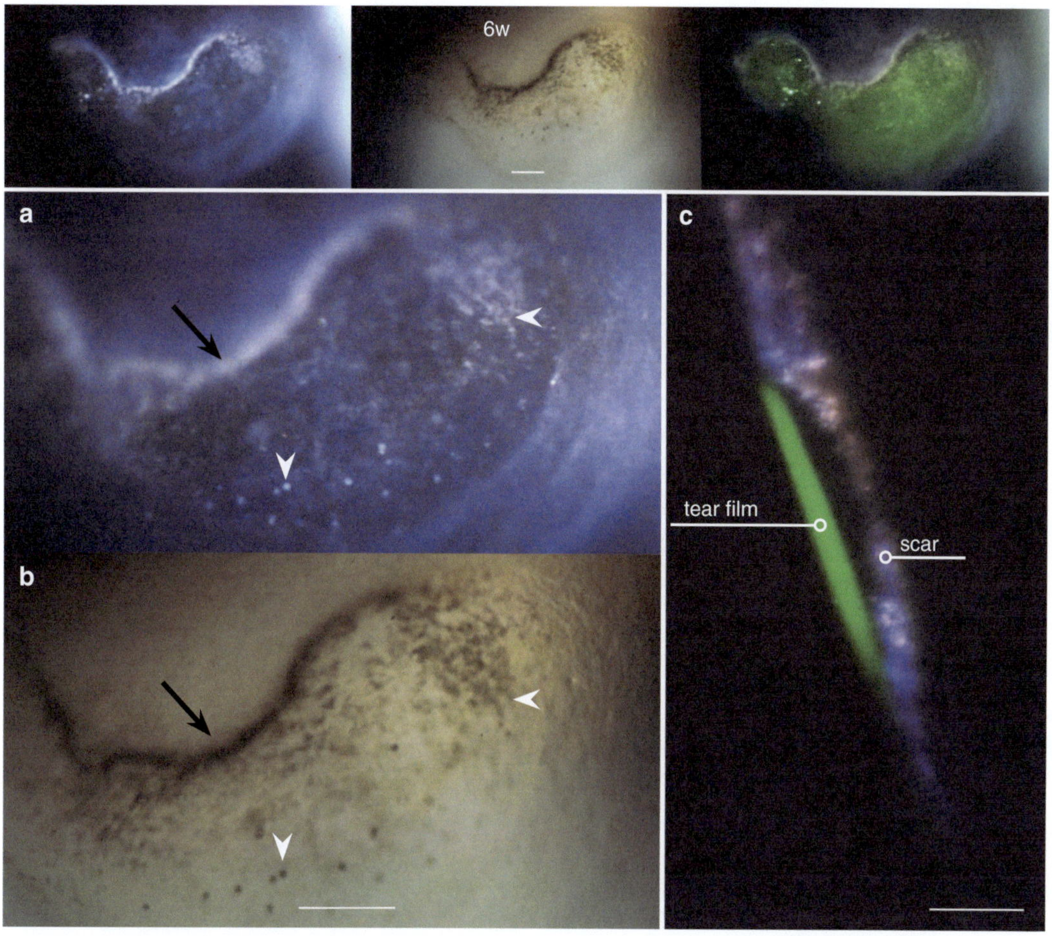

Fig. 3.8 *Upper row*: Survey of a kidney-shaped figure with light-reflecting periphery surrounding a relatively clear area in which (*right image*) fluorescein is pooling. (**a**, **b**) In the right part of the figure (*arrow*), many (**a**) light-reflecting individual or grouped dots (*arrowheads*), which (**b**) appear dark (*arrowheads*) in retroillumination, are visible. (The markers are placed in corresponding locations.) (**c**) shows the superficial location of the scar; the dark interval between the scar and the green stained tear film is presumably filled with epithelial cells

Comments

1. Although in this patient the fellow eye did not show any changes suggestive of epithelial dystrophy, the history of recurrent erosion implied the possibility of a sterile infiltrate; the *Staphylococcus aureus* found in the sample probably revealed lid colonisation in a patient with chronic blepharitis
2. In the case of stromal infiltrates the light-reflecting/dark dots, to start with abundantly present but successively disappearing, imply invading inflammatory cells

Case 3. Recurrent Stromal Infiltrates

Case Report

A 51-year-old healthy man with an 11-month history of spontaneous right-sided recurrent erosions presented with painful right eye. The paracentral lower cornea showed an edematous elevated area and the left one fingerprint lines and several cysts. Two days later, the right cornea showed a large erosion localised below the centre and generalised epithelial edema; the erosion healed within 4 days, but a superficial whitish stromal infiltrate, accompanied by anterior uveitis, developed in the paracentral lower cornea. Topical antibiotic had no effect, and microbiologic work-up revealed no pathogen. With dexamethasone 0.1 % eye drops twice a day, slowly tapered and discontinued after a month, the infiltrate resolved leaving a superficial scar. At that occasion, the epithelium showed two small elevations close to the scar, and approximately in that location, within 9 days, developed a new infiltrate. A second microbiologic work-up was also negative. He was again treated with topical steroid, very slowly tapered. When last seen, 2 months after the steroid was discontinued, the cornea showed two superficial scars. At both occasions were additionally noted small superficial opacities of uncertain origin, spread over the right cornea. Low Schirmer 1 test values implied dry eye.

The photographic series capturing the right cornea shows the residual scar after the first episode and changes occurring during the second episode between day 1 (3 days of steroid treatment) and 4, 5 months.

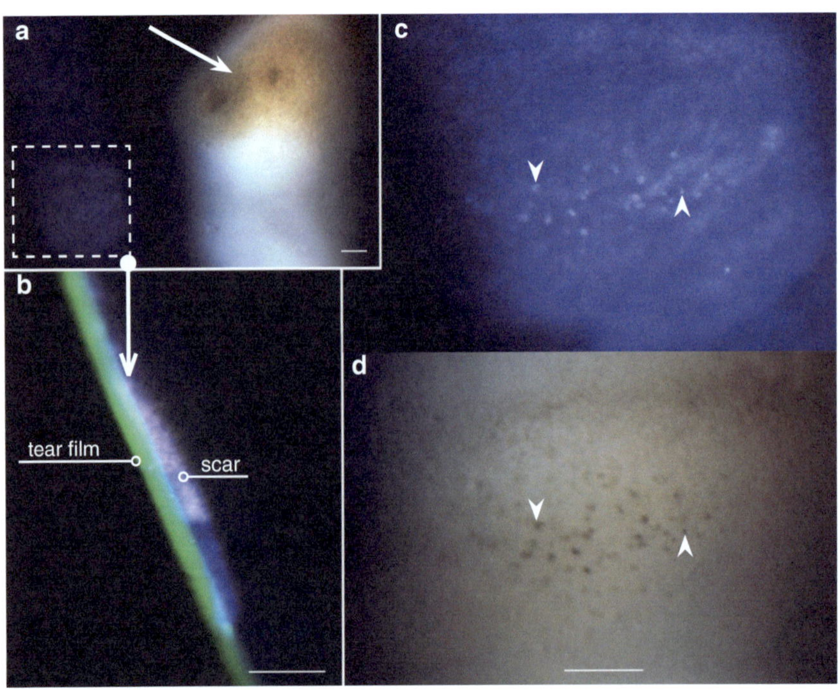

Fig. 3.9 Scarring after the first corneal infiltrate about 7 weeks after onset. (**a**) Survey capturing the scar (in *frame*); to the right a new infiltrate(s) (*arrow*) on day 1 is visible. (**b**) shows the superficial position of the scar below tear film stained green with fluorescein. (**c**, **d**) The area of the scar shows larger and smaller dots (*arrowheads*), (**c**) light-reflecting and (**d**) appearing dark in retroillumination. (The arrowheads are placed in corresponding locations.) The infiltrate(s) indicated in (**a**) by *arrow* is shown in Figs. 3.10, 3.11, 3.12, 3.13 and 3.14 ((**b**) Adapted from Tabery (1998))

Case 3. Recurrent Stromal Infiltrates (cont.)

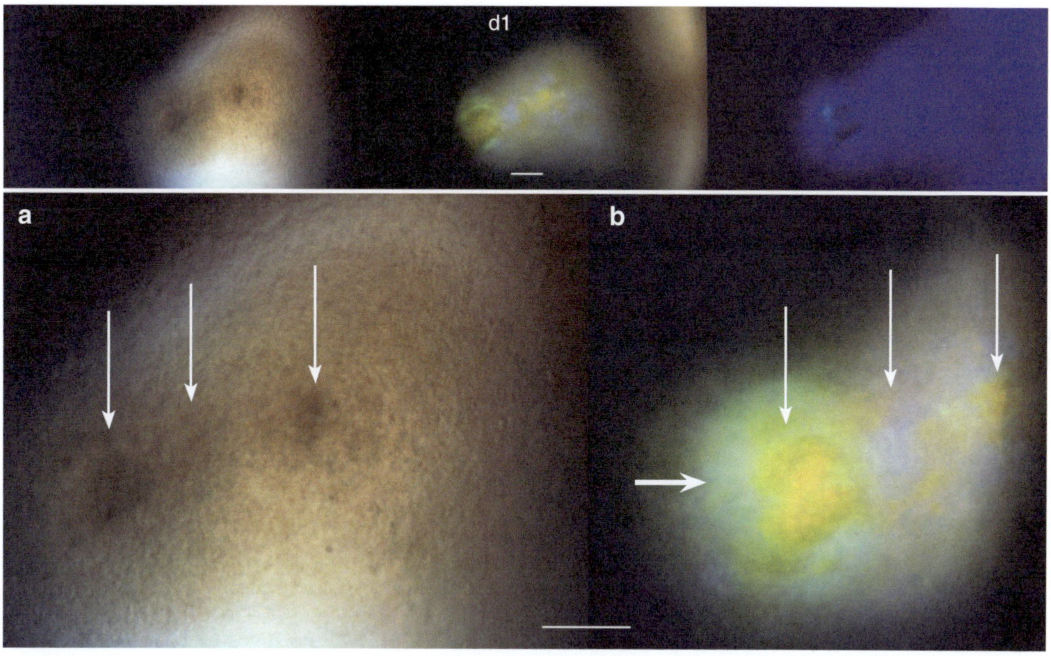

Fig. 3.10 *Upper row*: Survey of seemingly 2 rounded infiltrates floating together, captured before (*left*) and after staining with fluorescein (centre and right). (**a**) Denser spots (*arrows*) within the infiltrate(s) which (**b**) stain yellow with fluorescein; the *short arrow* indicates surrounding green fluorescein staining; cf. upper row, right (blue filter) (the long arrows are placed in corresponding locations)

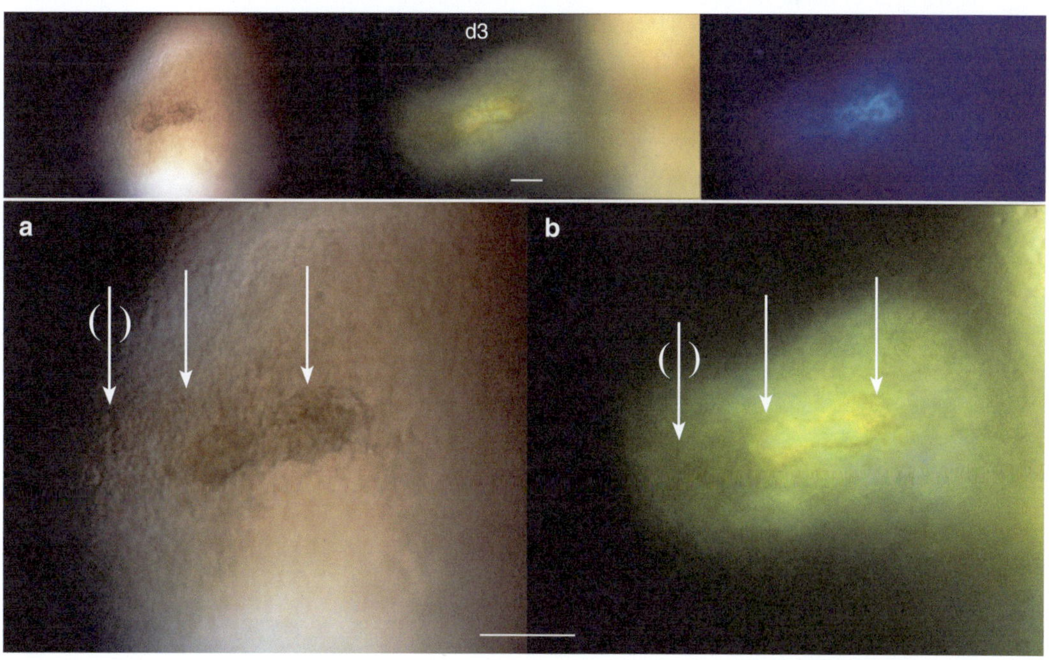

Fig. 3.11 *Upper row*: Survey captured before (left) and after staining with fluorescein (centre and right). (**a**) 2 denser spots (*arrows*) in apposition. The left one indicated in Fig. 3.10a seems to have partly cleared. (**b**) Yellow fluorescein staining within the area of the denser spots (*arrows*); green fluorescein staining is better visible with the blue filter (*upper row*, right) (the arrows are placed in locations approximately corresponding to those in Fig. 3.10)

Case 3. Recurrent Stromal Infiltrates (cont.)

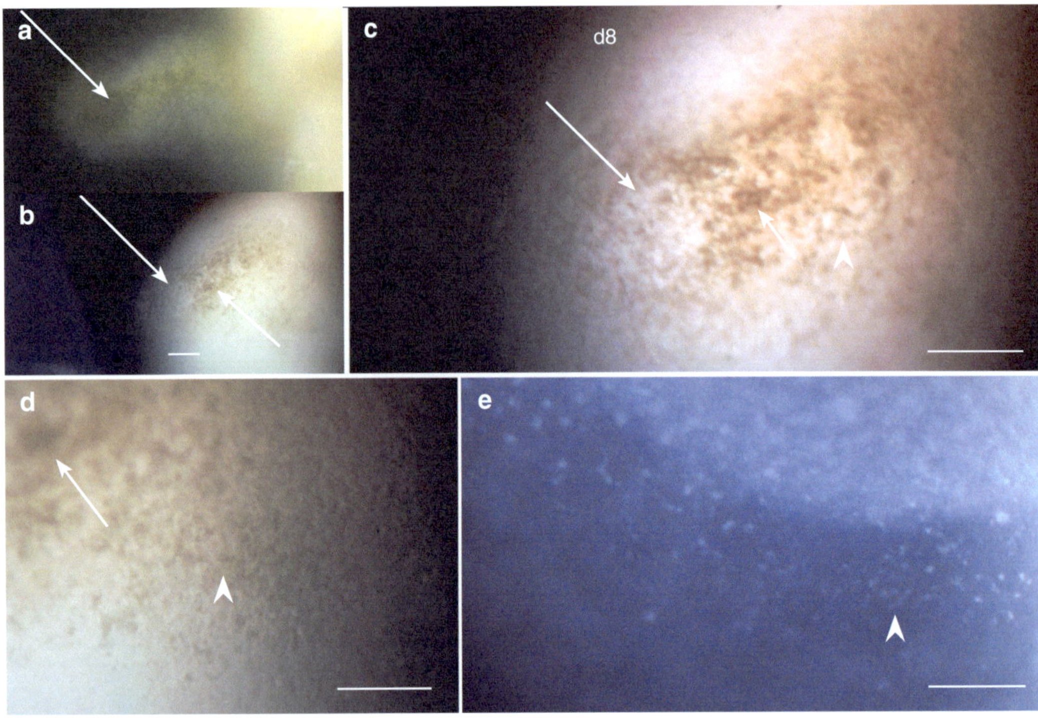

Fig. 3.12 (**a**, **b**) Survey. The central part of the infiltrate(s) (*long arrow*) is clearing; the *short arrow* indicates the same location as in (**c**) and (**d**). (**c**) The area indicated by *long arrow* appears brighter than the adjacent one that contains many dark dots, individual (*arrowhead*) and grouped (*short arrow*). (**d**) Dots (*arrowhead*) in the periphery of the infiltrate(s); the *arrow* indicates the same location as in (**a**–**c**). (**e**) Light-reflecting dots (*arrowhead*) in the periphery of the infiltrate(s)

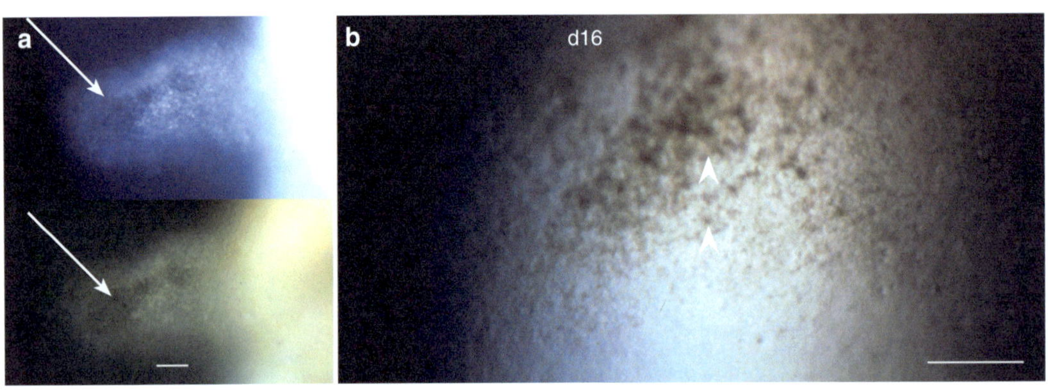

Fig. 3.13 (**a**) Survey. Ongoing clearing of the central part (*arrows*) of the infiltrate(s). (**b**) Dots (*arrowheads*), individual and grouped, in the periphery of the infiltrate(s)

Case 3. Recurrent Stromal Infiltrates (cont.)

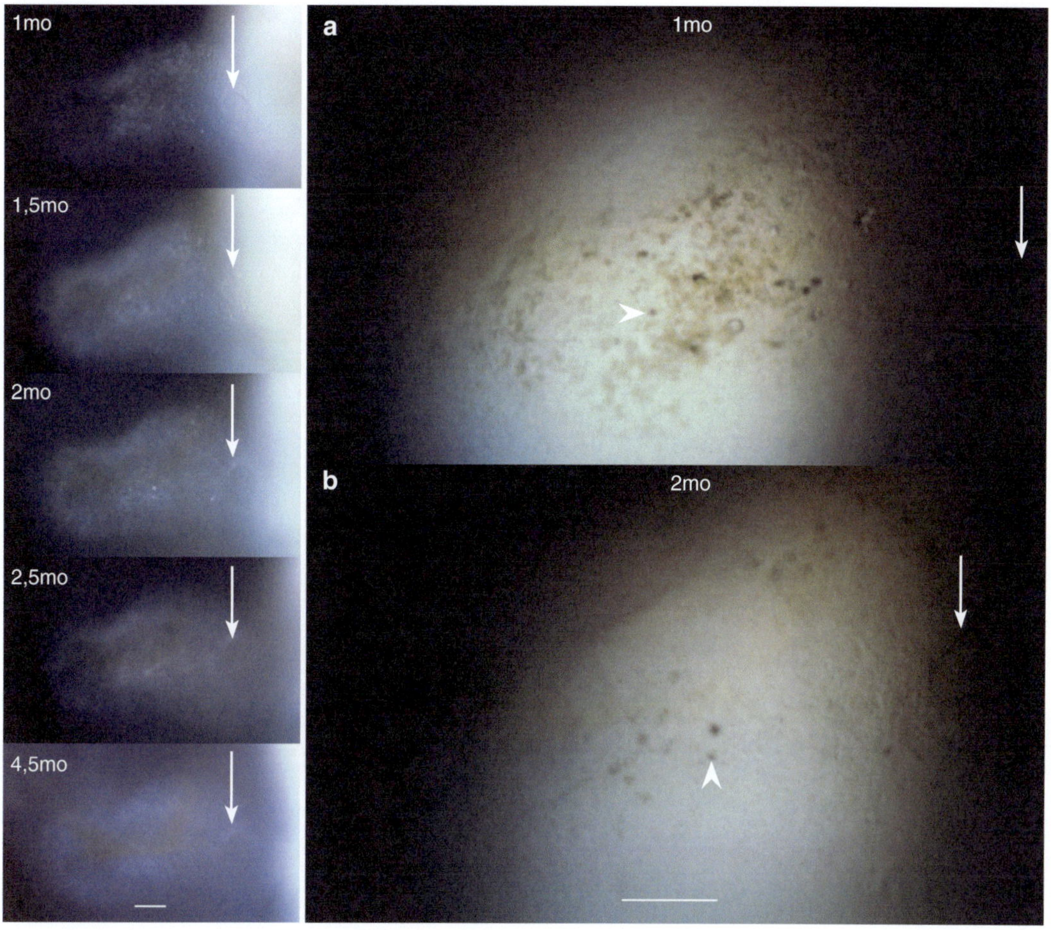

Fig. 3.14 The development in a subepithelial scar. For orientation, the *arrows* indicate the same location in all photographs. *Left column*: Survey. The size of the scar remains more or less the same. (**a, b**) The numbers of the dots (*arrowheads*) successively diminish

Comments

1. In this patient, the history of recurrent erosion in conjunction with epithelial dystrophic changes in the fellow eye was suggestive of a sterile infiltrate
2. In the case of stromal infiltrates, the light-reflecting/dark dots, to start with abundantly present but successively disappearing, imply invading inflammatory cells

Part II

Corneal Epithelial Edema

About Corneal Epithelial Edema

Corneal epithelial edema results from excess water in the epithelium. It occurs in conditions compromising the function of the endothelial layer such as corneal dystrophies (of which Fuchs' dystrophy is the most common one), intraocular inflammation, and high intraocular pressure; also, the endothelium may be damaged by trauma or during ocular surgery. Another cause of epithelial edema may be improper use of contact lenses. Epithelial edema causes deterioration of vision, haloes around lights, photophobia, ocular discomfort, foreign body sensation, and pain.

4 The Morphology of Corneal Epithelial Edema

There are two manifestations of corneal epithelial edema, an intracellular and an extracellular one.

Intracellular edema shows as a fine granular pattern caused by swollen cells of about 10–12 μm in diameter. *Extracellular* fluid accumulation results in surface elevations appearing partly as variously large (about 20–90 μm), well-defined, transparent *blisters* and partly as large *bullae* (the here shown ones measuring about 350 μm to >1 mm in diameter). The bullae, usually situated in areas showing intracellular edema and blisters, are rounded, oval or sometimes geographic or irregular but always well delineated; many show a surrounding blister-free rim about 50 (–80) μm broad. Their contents are clear. Occasionally are also present small protrusions (of about 22–25 μm in diameter) that might represent either swollen suprabasal cells or incipient blisters.

With *fluorescein*, blisters appear either as dark protrusions in the green stained tear film or rounded brilliantly green spots; intact bullae do not stain.

Some corneae show *additional features*: variously shaped strongly light-reflecting figures with granular contents ('opaque cysts'), a group of phenomena such as fusiform expansions or pear-shaped cystic spaces often arranged in rows and containing rounded bodies and epithelial lines interconnecting cystic spaces or implying advancing edges of epithelial sheets. All these features are strikingly similar to or indistinguishable from those occurring in the recurrent erosion syndrome and/or BMD (cf. Chap. 1).

In chronic corneal edema additional changes such as *epithelial wrinkles*, often appearing as roughly parallel superficial lines, *invading vessels* and *subepithelial fibrosis* may develop.

H.M. Tabery, *Recurrent Erosion Syndrome and Epithelial Edema: In Vivo Morphology in the Human Cornea*,
DOI 10.1007/978-3-319-06545-8_4,

Intracellular Edema and Small Bullae

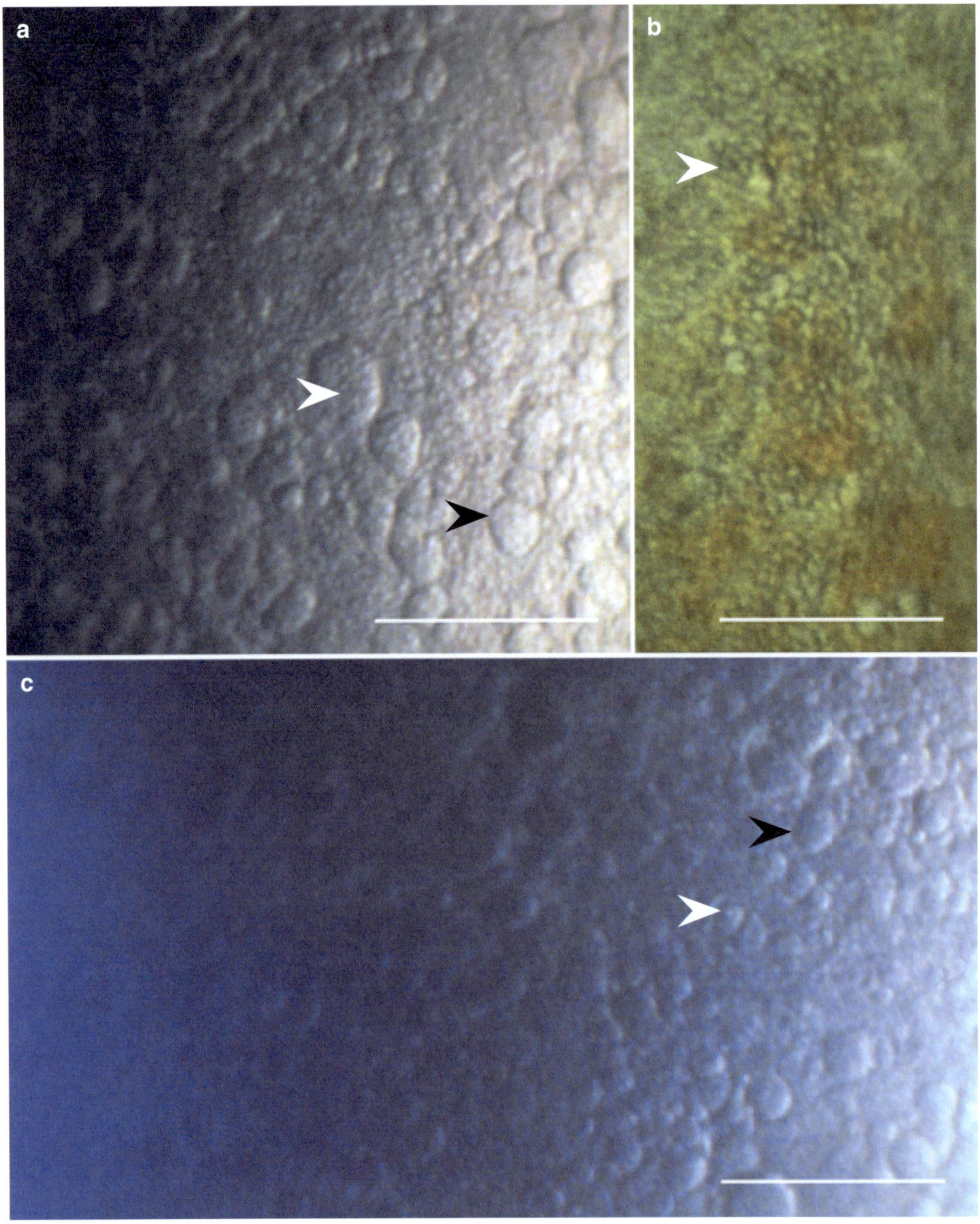

Fig. 4.1 (a–c) Fine granular pattern of intracellular edema (*white arrowheads*) and (a, c) variously large blisters (*black arrowheads*). The yellow-brown spots in (b) are caused by fluorescein pooling presumably within blisters (cf. Fig. 4.2)

Fluorescein Staining (1)

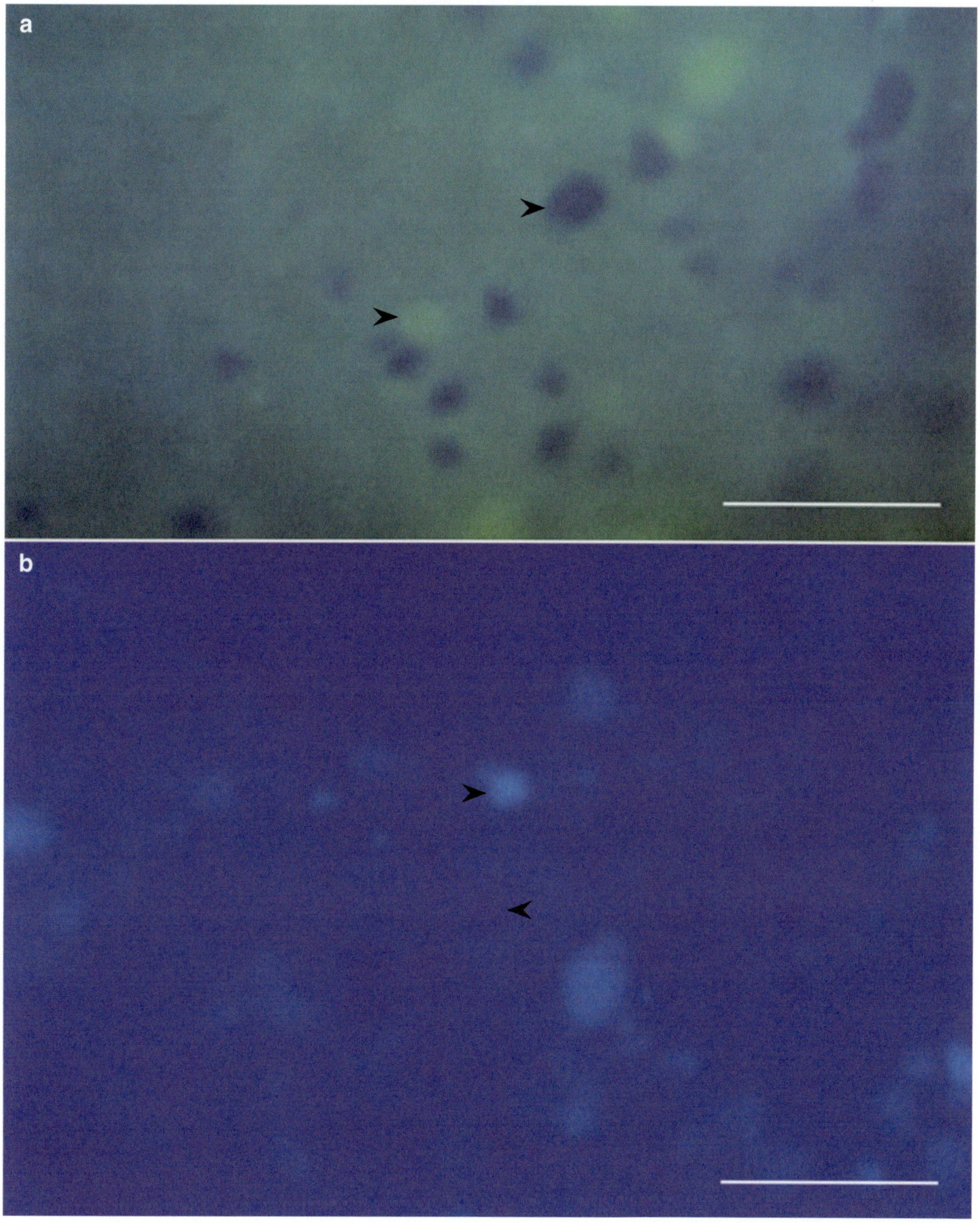

Fig. 4.2 (**a**, **b**) With fluorescein, some blisters (*arrowheads*) stain brilliantly green while others appear dark in the green stained tear film (**b**, blue filter)

Fluorescein Staining (2)

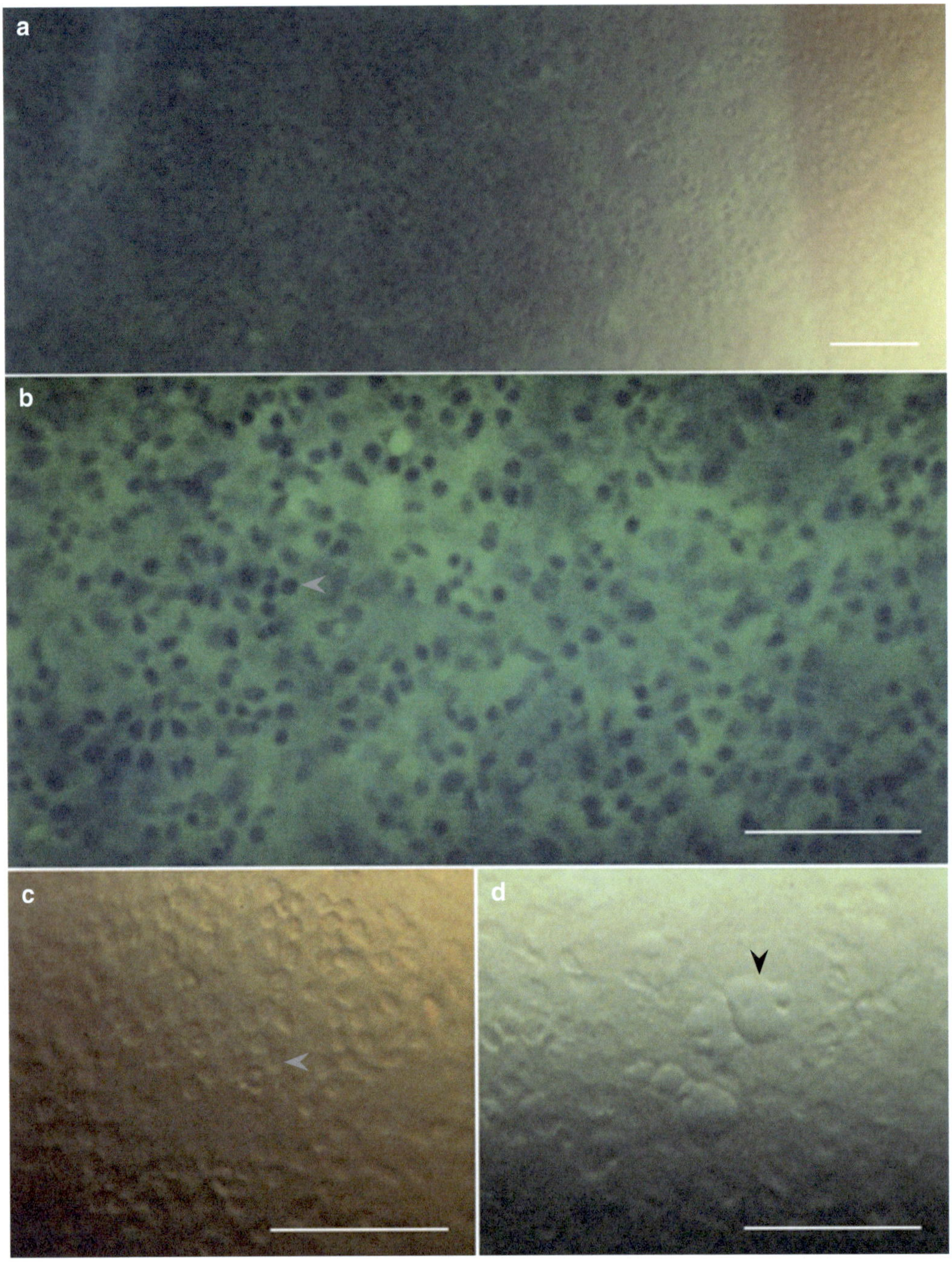

Fig. 4.3 (a) Survey of an area showing (b) dot-like structures (*arrowhead*) protruding in the green stained tear film; these protrusions correspond to (c) small surface excrescences (*arrowhead*). (d) In a different area of the same cornea, variously large blisters are visible (*arrowhead*) (decompensated cornea in a prephtitic eye)

Comment

The small excrescences imply swollen suprabasal cells or incipient blisters.

Fluorescein Staining (3)

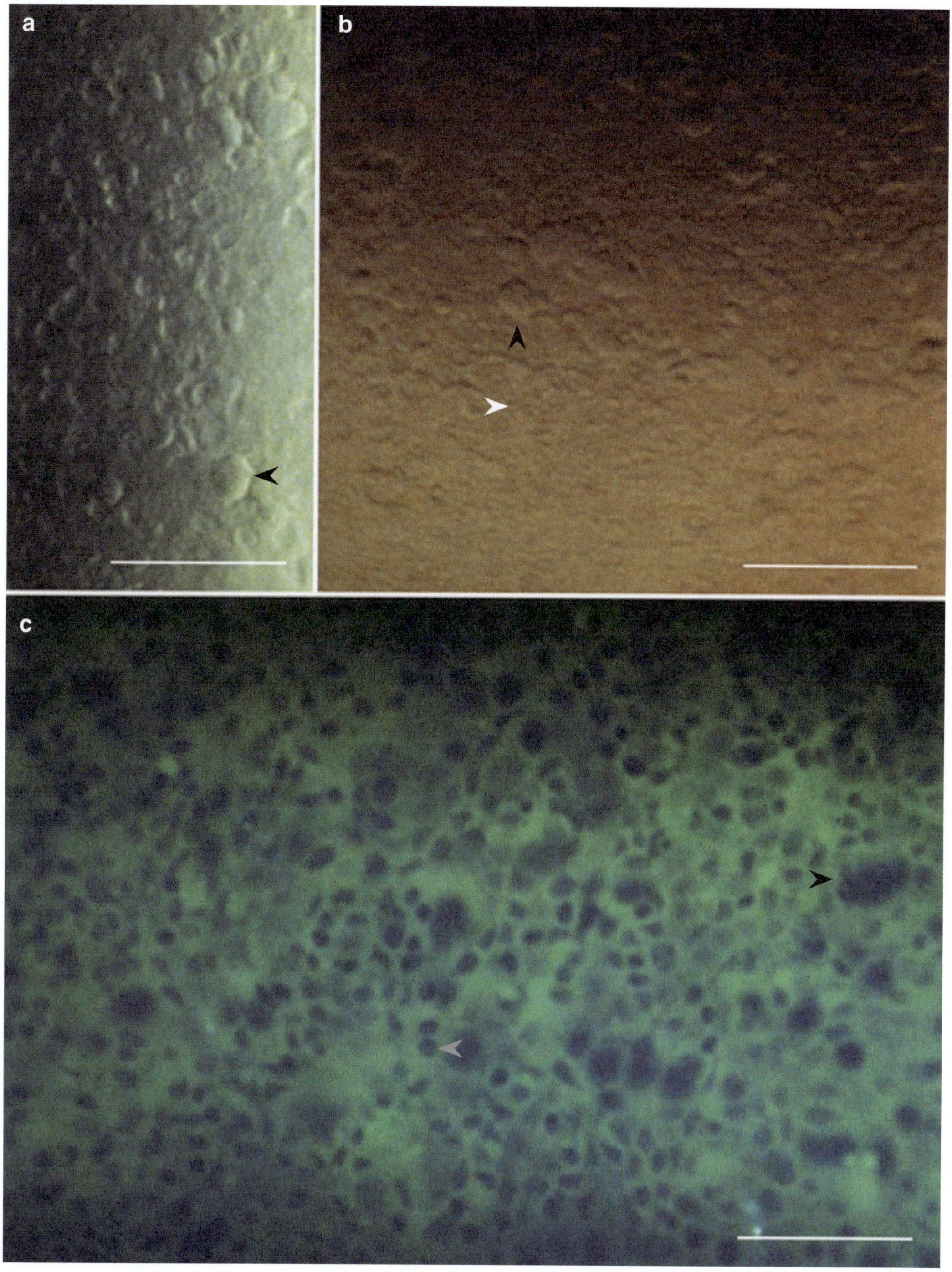

Fig. 4.4 (**a**, **b**) Two areas showing variously large blisters (*black arrowheads*) and (**b**) a fine granular pattern of intracellular edema (*white arrowhead*). (**c**) In this area, with fluorescein, small protrusions (*grey arrowhead*) similar to those shown in Fig. 4.3 and larger ones (*black arrowhead*) implying blisters are visible (the same cornea as in Fig. 4.3; cf. also Fig. 4.12d)

Bullous Epithelial Detachment

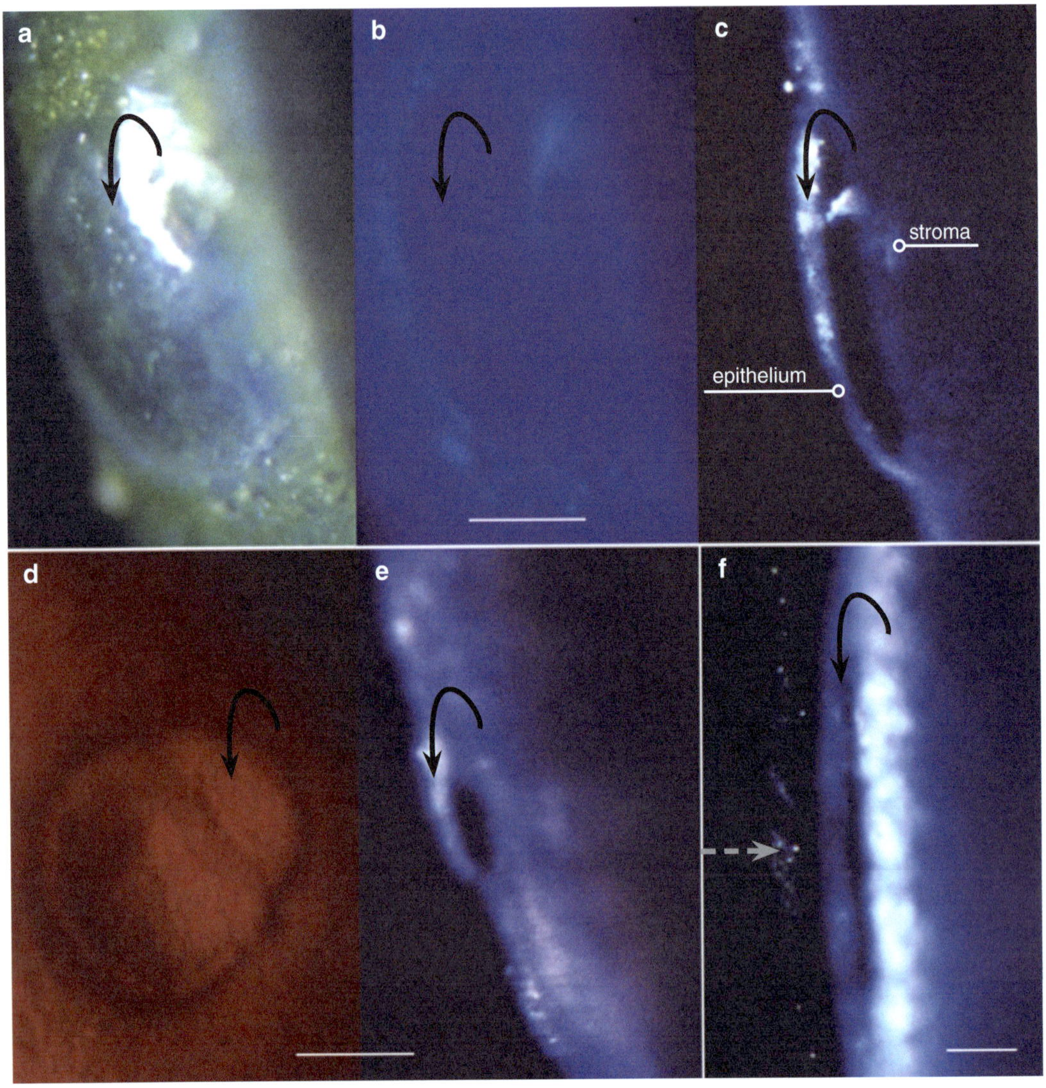

Fig. 4.5 (**a–c**) A large bulla (*arrows*) that (**a**, **b**) does not stain with fluorescein. (**c**) shows the detached epithelium (*arrow*) in a narrow light beam; the space between the epithelium and the stroma is filled with clear fluid. ((**b**) blue filter.) (**d–e**) A smaller bulla (*arrows*) captured in (**d**) retroillumination and (**e**) a narrow light beam creating optical section through its periphery. (**f**) Optical section through a flat bulla (*bowed arrow*) situated below a contact lens (*dashed arrow*) (see also Figs. 5.1, 5.5 and 5.6)

Opaque Cysts, Epithelial Wrinkles and Subepithelial Fibrosis

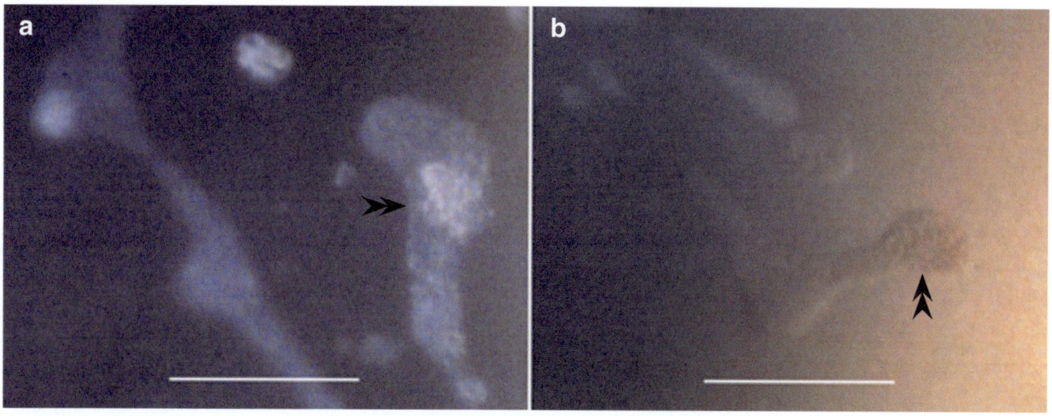

Fig. 4.6 Variously shaped, well delineated and (**a**) more or less strongly light-reflecting opaque cysts (*arrowhead*) containing (**b**) granular material (*arrowhead*) (cf. Figs. 1.6 and 1.7 and Chap. 5, Cases 2 and 3)

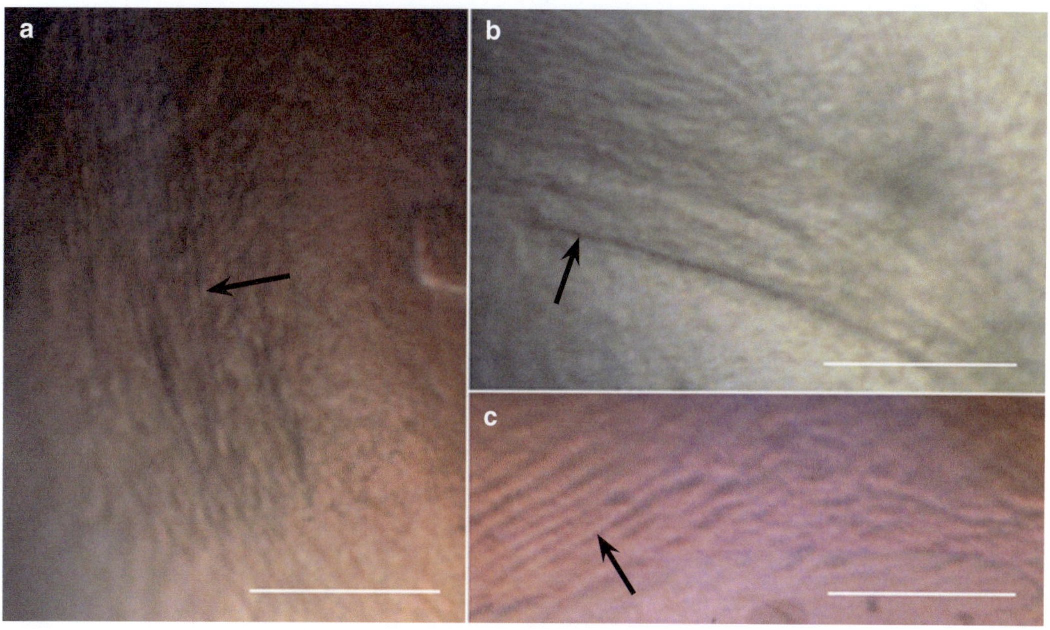

Fig. 4.7 (**a**–**c**) Epithelial wrinkles (*arrows*) (see also Fig. 5.8)

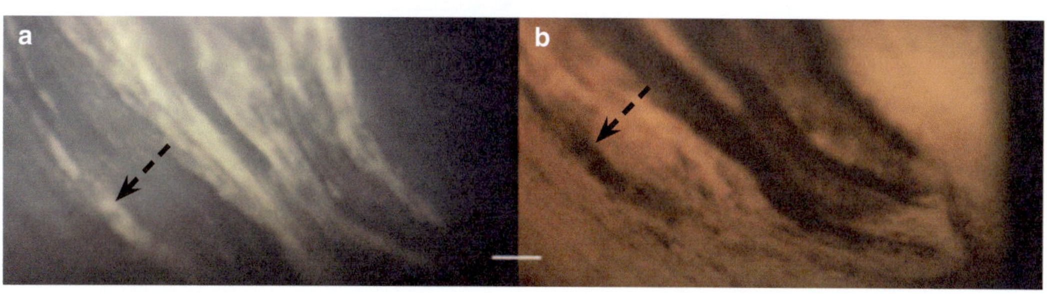

Fig. 4.8 Subepithelial fibrosis. This area shows (**a**) strongly light-reflecting and (**b**) optically dense tissue (*arrows*), with the slit lamp judged as situated below the epithelium

Other Cystic Spaces (1)

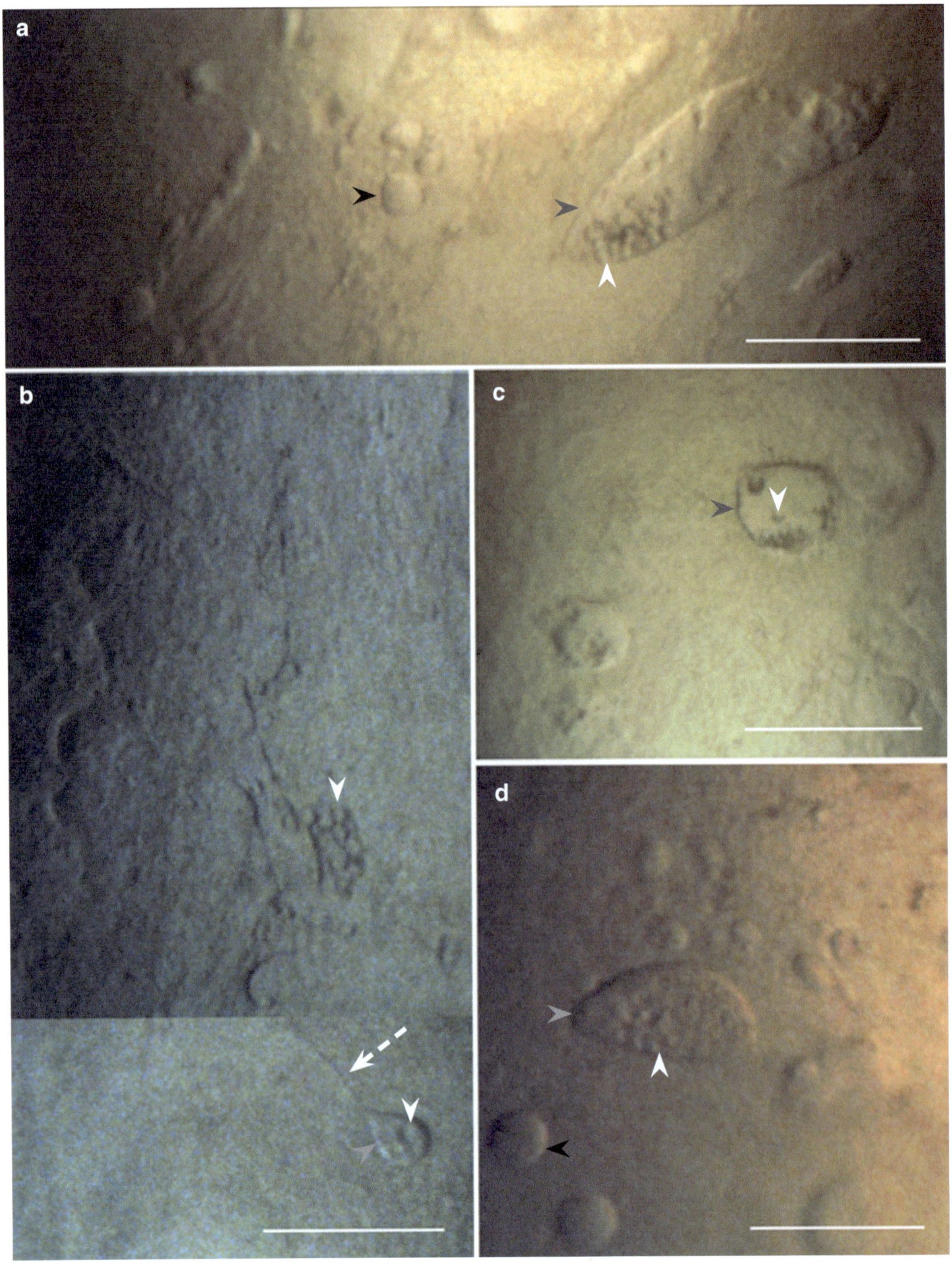

Fig. 4.9 (**a–d**) Variously shaped and variously large cystic spaces (*grey arrowheads*) containing rounded bodies (*white arrowheads*), rounded cystic spaces/blisters (**a**, **d**, *black arrowheads*) and (**b**) an epithelial line (*arrow*), captured in 4 corneae with chronic edema (cf. Fig. 4.10)

Other Cystic Spaces (2)

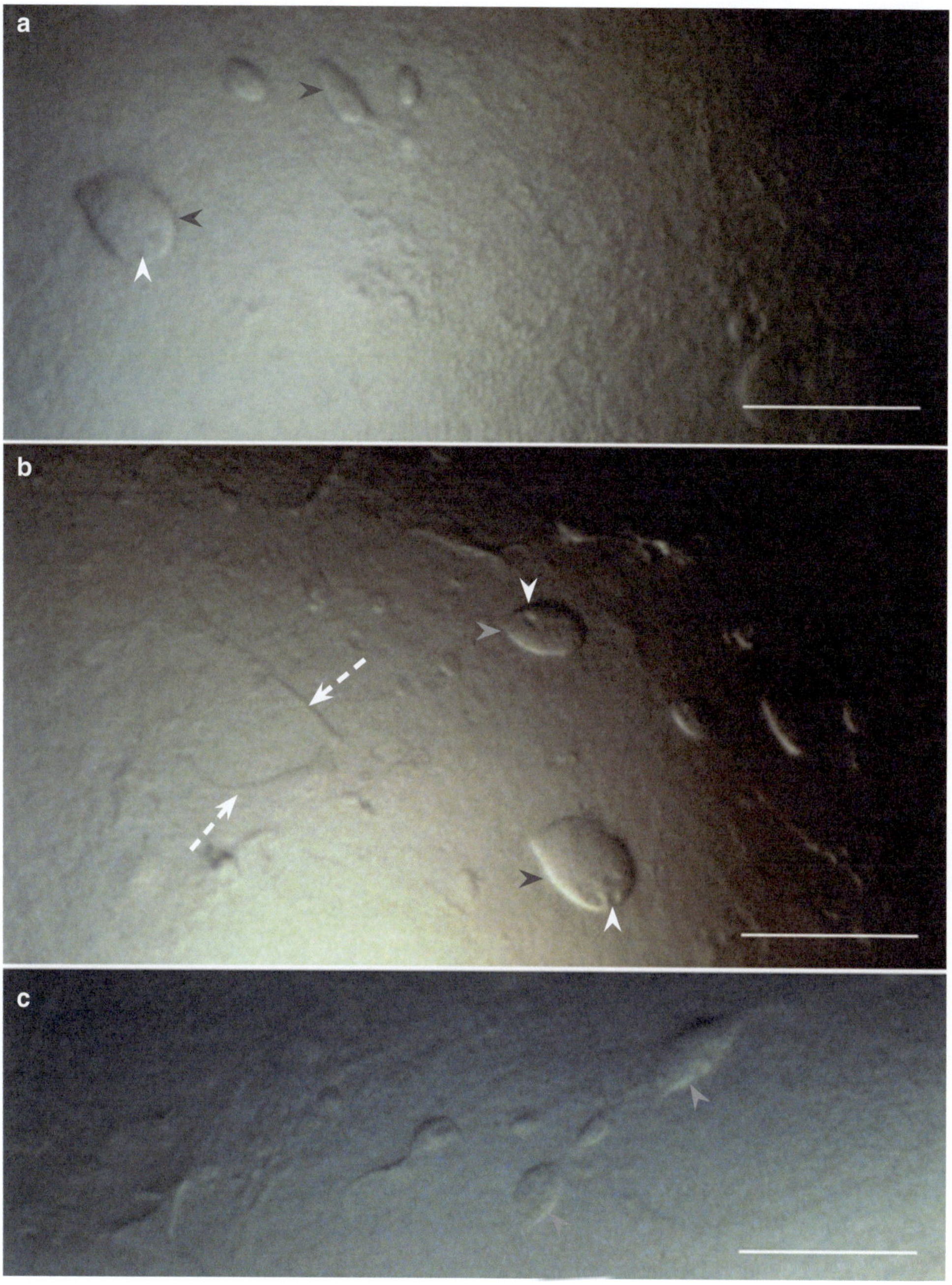

Fig. 4.10 (**a–c**) Variously shaped and variously large cystic spaces (*grey arrowheads*) containing rounded bodies (*white arrowheads*) and (**b**) epithelial lines (*arrows*), all captured in a cornea during an attack of acute glaucoma. In (**a**), to the right, a fine granular pattern of intracellular edema is additionally visible

Compare Fig. 4.12 and Chap. 5, Case 1

Comment

Figures 4.9 and 4.10 strongly resemble epithelial changes captured in recurrent erosion syndrome (Part 1).

Cornea Guttata (1)

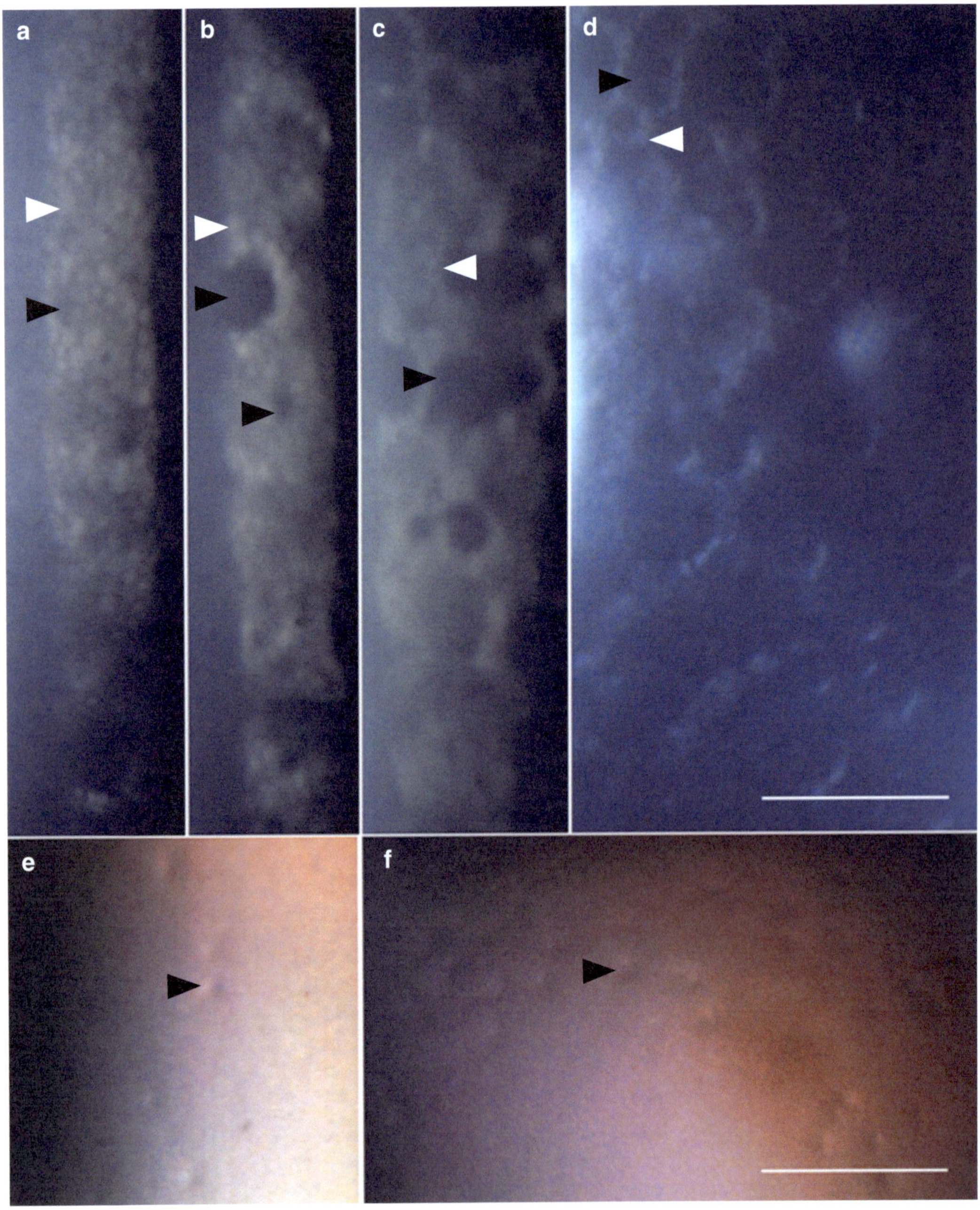

Fig. 4.11 Cornea guttata. (**a–d**) In specular reflection light-reflecting endothelial cells (*white arrowheads*) and variously large dark areas (*black arrowheads*) representing subendothelial excrescences (guttae) are visible. (**e**, **f**) In retroillumination, small guttae (*black arrowheads*) appear as small endothelial protrusions ((**a–f**) show different areas of the same, not edematous, cornea)

Cornea Guttata (2)

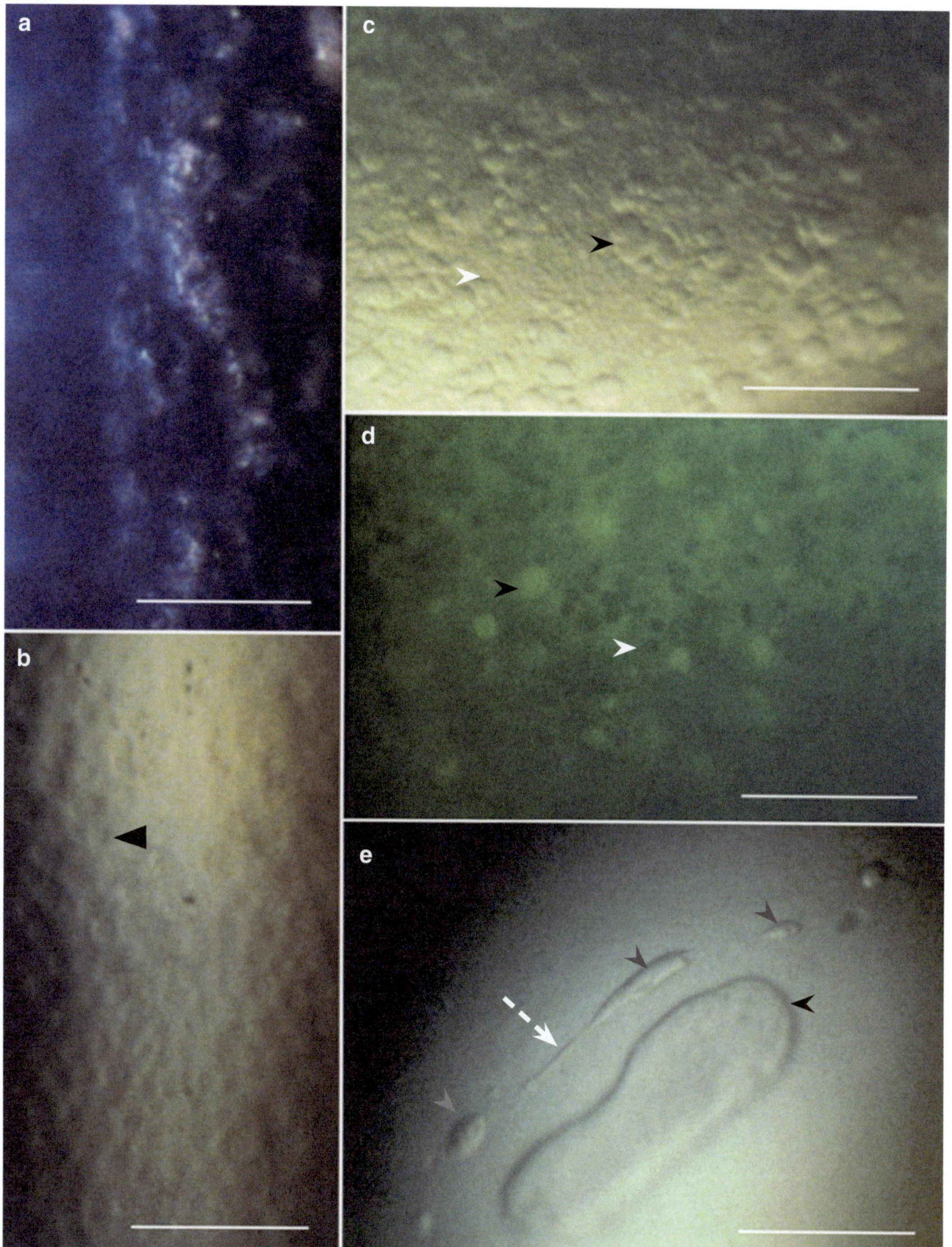

Fig. 4.12 In this area, (**a**) the endothelial specular reflection is strongly distorted. (**b**) shows an area containing many guttae (*arrowhead*); (**c**) epithelial blisters (*black arrowhead*) and a fine granular pattern of intracellular edema (*white arrowhead*); (**d**) blisters staining brilliantly green with fluorescein (*black arrowhead*) and small, dark protrusions (*grey arrowhead*, cf. Figs. 4.3 and 4.4); and (**e**) small cystic spaces (*grey arrowheads*) and an epithelial line (*arrow*) close to a larger cystic space (*black arrowhead*) (Fuchs' dystrophy)

Invading Vessels

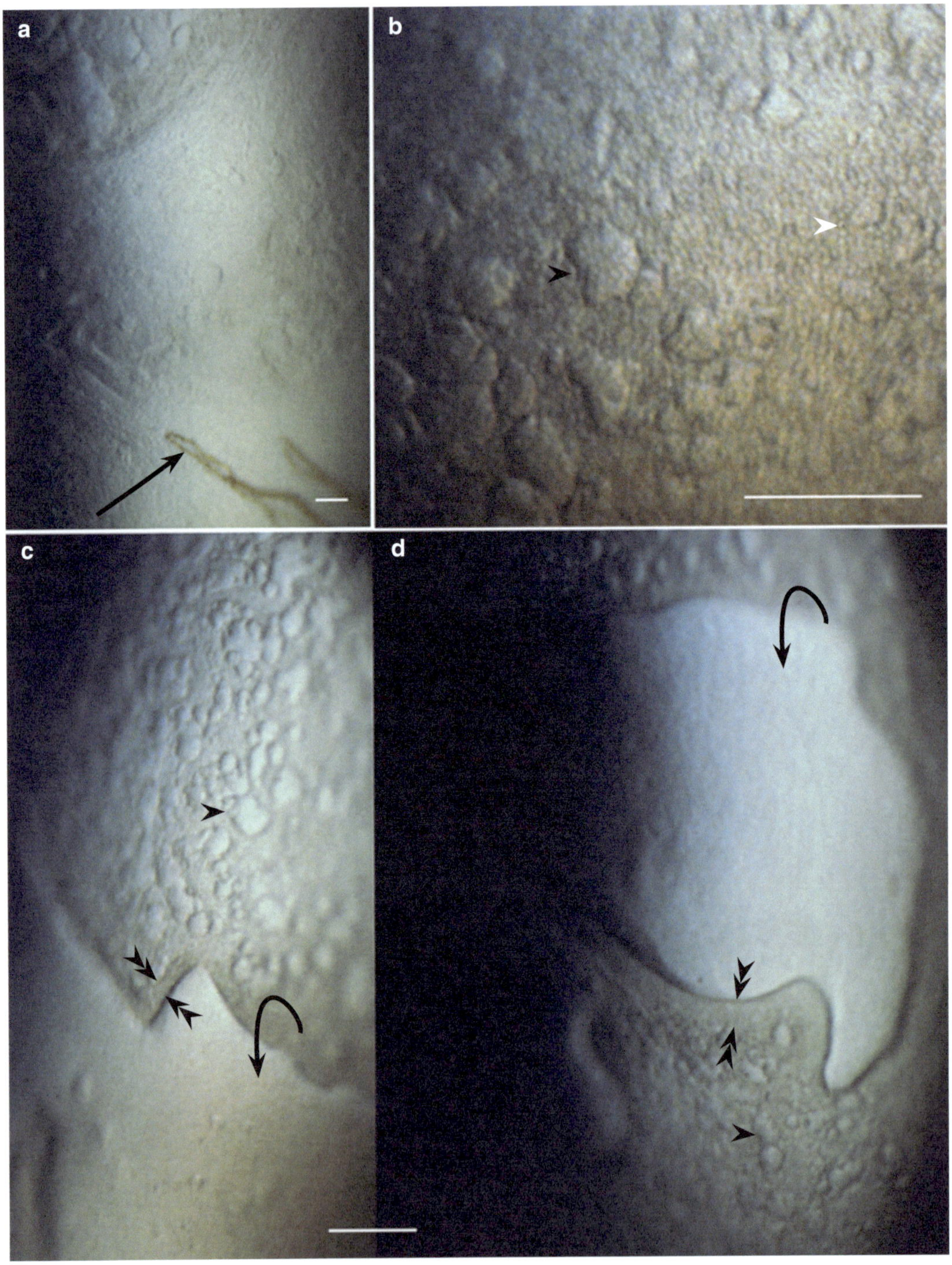

Fig. 4.13 (a) Invading vessels (*arrow*), (b) smaller and larger blisters (*black arrowhead*) and a fine granular pattern of intracellular edema (*white arrowhead*), (c) the left and (d) the right part of a large bulla (*arrow*) the edge of which shows a distinct rim (between the *double arrowheads*); in the surroundings are visible blisters (*black arrowheads*) (see also Figs. 5.1, 5.5 and 5.6 (chronic epithelial edema))

Filaments

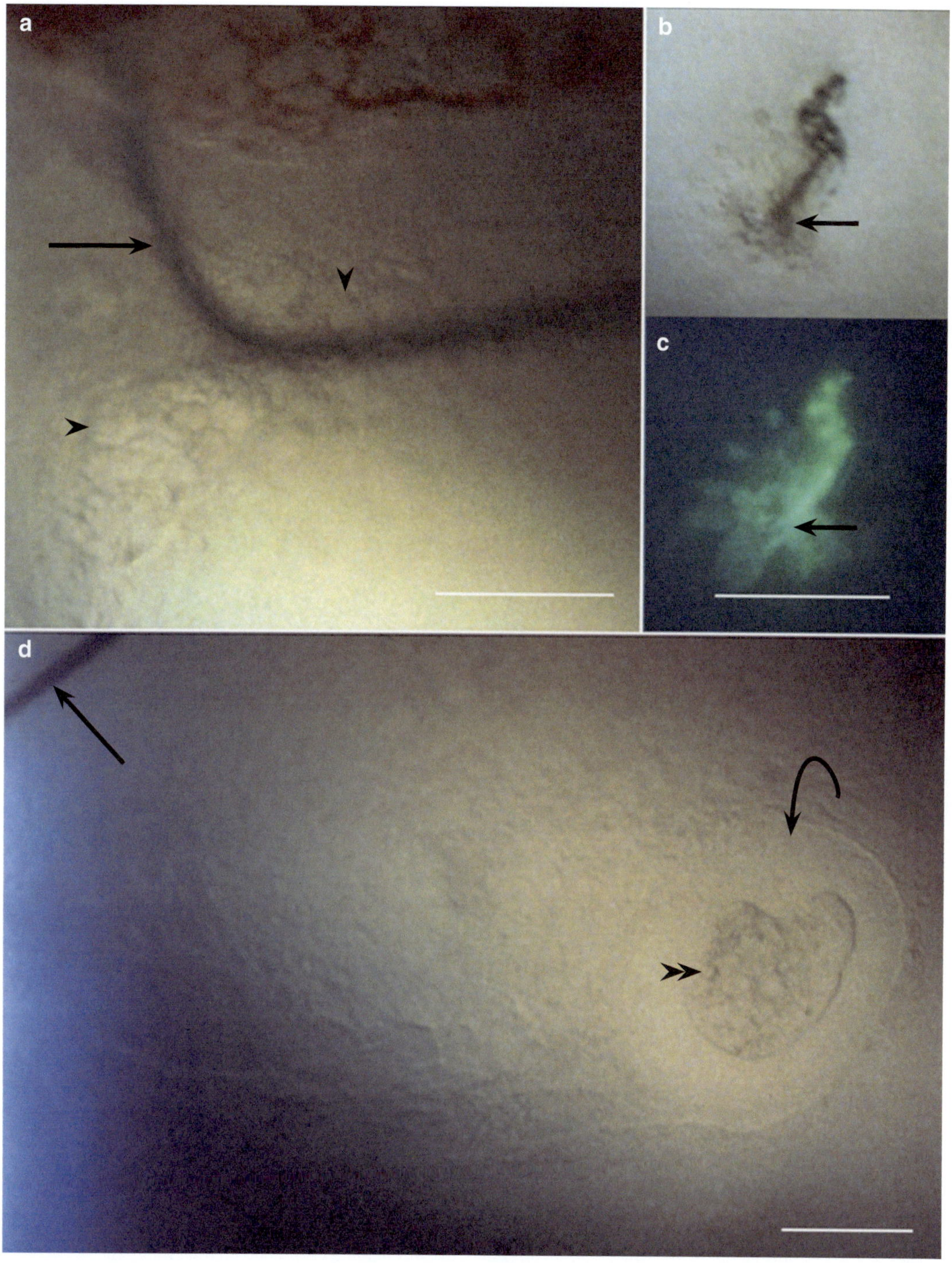

Fig. 4.14 (a) Variously large blisters (*arrowheads*) close to the site of superior incision in cataract surgery; the *arrow* indicates a suture left in situ. (b, c) Remnant of a filament (*arrow*) present close to the incision site. (d) Close to the suture (*straight arrow*) an area (*bowed arrow*) containing a denser (darker) patch of granular material (*arrowheads*) is visible (cf. Fig. 4.15, overleaf)

Combination of Features (1)

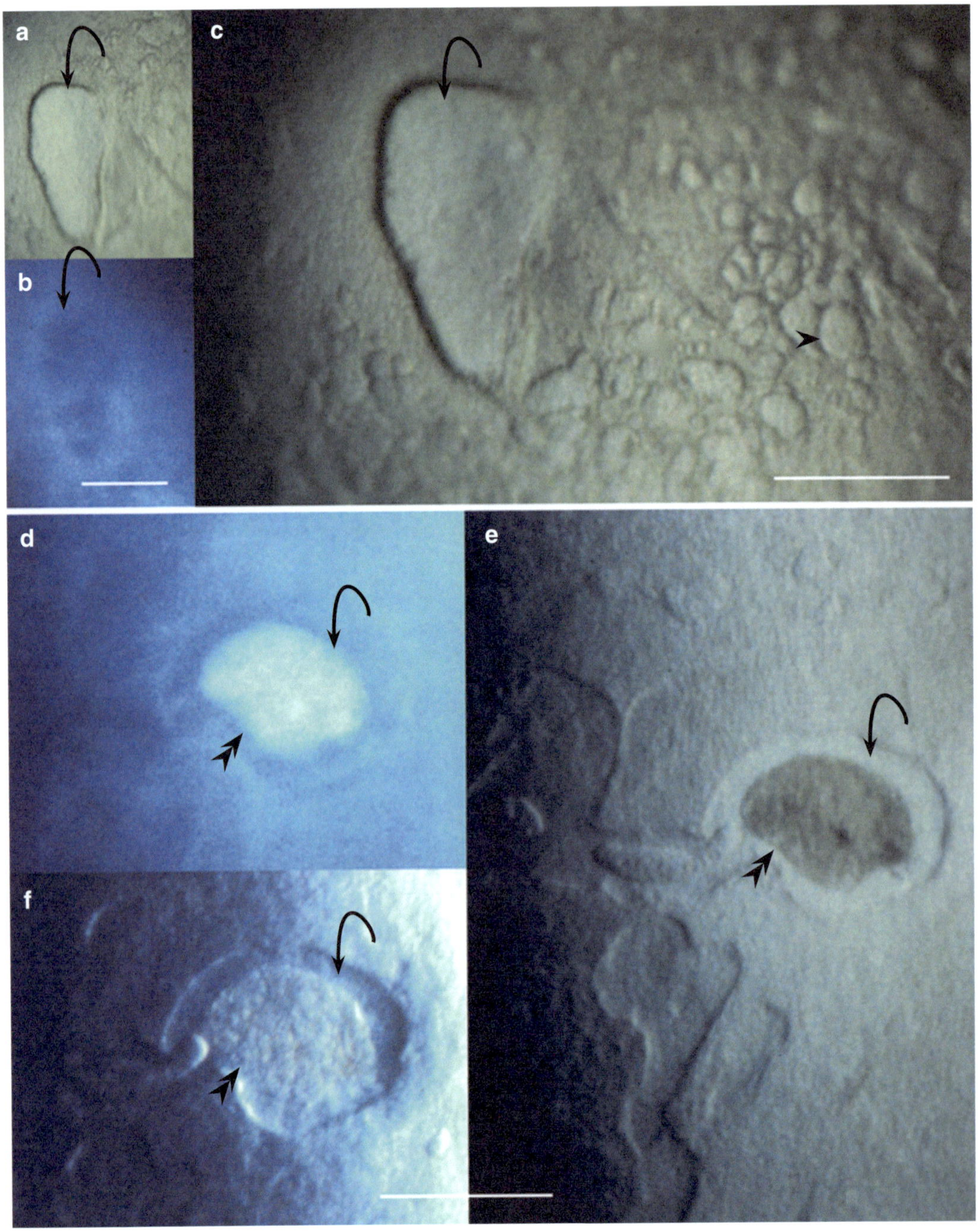

Fig. 4.15 (**a–c**) A bulla (*arrows*) with (**b**) clear (dark) contents (**c**) the surroundings of which show blisters (*arrowhead*) and a fine granular pattern of intracellular edema. (**d–f**) A patch of (**d**) strongly light-reflecting, (**e**) dense (dark) and (**f**) granular material (*arrowheads*); the surrounding clear zone (*arrow*) implying the presence of clear fluid is suggestive of a cystic space (the markers are placed in corresponding locations)

(Lower part of the right cornea of a 33-year-old woman with Marfan's syndrome, bilateral cataract surgery and iris-fixated anterior chamber lenses 13 years previously. The left cornea showed similar changes)

Combination of Features (2)

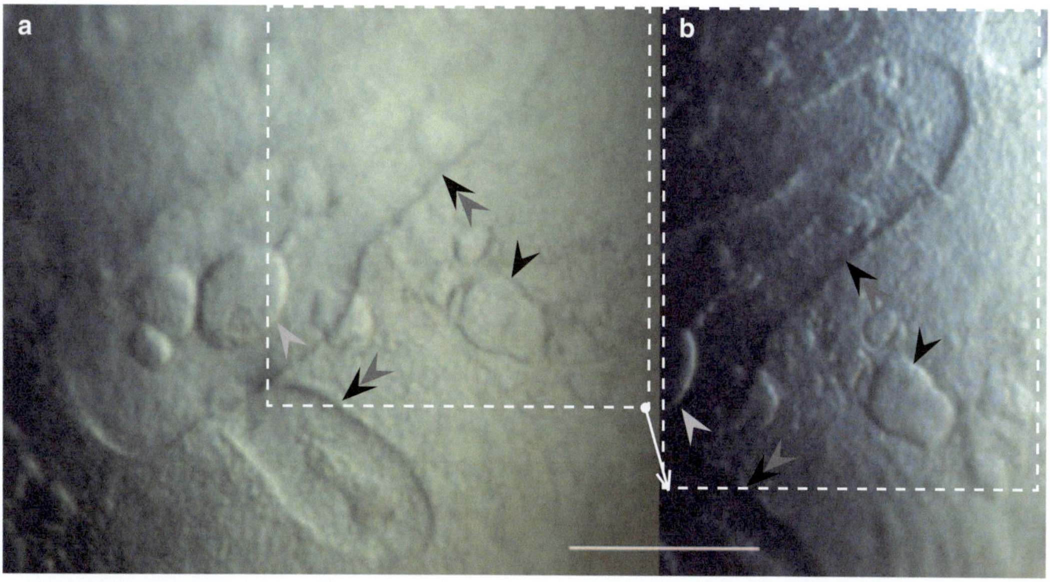

Fig. 4.16 (a) In this area of the cornea shown in Fig. 4.15 rounded cystic spaces (*grey arrowhead*), blisters (*black arrowhead*) and a fine granular pattern of intracellular edema are visible; the lines indicated by *double arrowheads* might represent edges of larger cystic spaces. The *frame* indicates the part of this area that is in focus in (b); (the arrowheads are placed in corresponding locations)

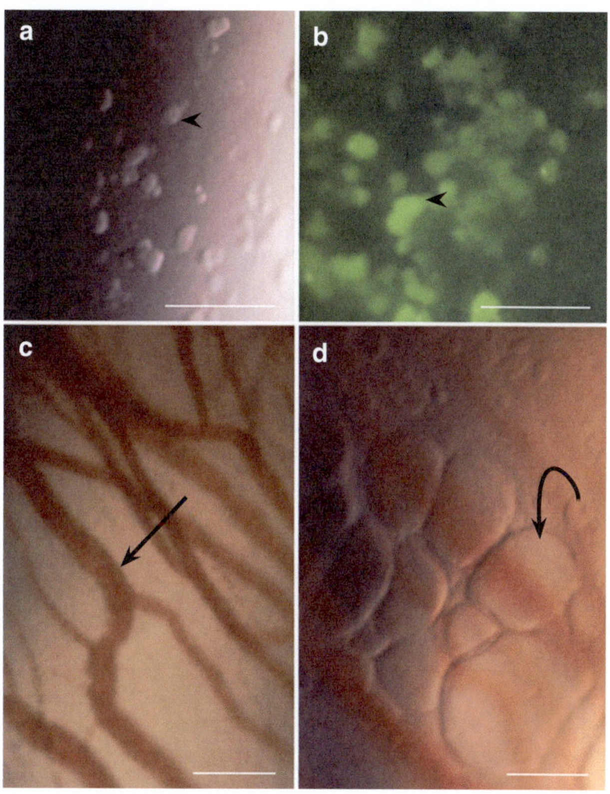

Fig. 4.17 (a) Blisters (*arrowhead*) staining (b) brilliantly green with fluorescein (*arrowhead*), (c) invading vessels (*arrow*) and (d) bullae (*arrow*) (severe keratomalacia in vitamin A deficiency)

5 Case Reports

Corneae with epithelial edema occasionally show epithelial phenomena similar to or indistinguishable from those occurring in the recurrent erosion syndrome and/or EBMD. The photographic sequences presented in this chapter visualise some of their dynamics.

Of the 3 patients, epithelial edema was documented in Case 1 (Fuchs' dystrophy) and Case 2 (relatable to intraocular inflammation). In Case 3 (with bilateral cornea guttata and cataract surgery with superior corneal incision in the affected eye) it is probable that epithelial edema had occurred close to the incision site.

In Case 1 features compatible with ongoing epithelial activity, in the recurrent erosion syndrome interpreted as signs of repair, were repeatedly captured; Cases 2 and 3 show the dynamics of light-reflecting intraepithelial granular material optically indistinguishable from opaque cysts ('dots') in the recurrent erosion syndrome and/or EBMD. (See also Final Remark.)

H.M. Tabery, *Recurrent Erosion Syndrome and Epithelial Edema: In Vivo Morphology in the Human Cornea*,
DOI 10.1007/978-3-319-06545-8_5,

Case 1. Epithelial Activity in Chronic Edema

Case Report

A 78-year-old man with a history of right-sided acute glaucoma and laser iridectomy in both eyes 10 years previously presented with irritation, pain and worsening vision in the right eye for about 10 days. The right eye showed central corneal epithelial edema and within it cystic changes reminiscent of recurrent erosion syndrome, stromal edema, Descemet's membrane folds, cornea guttata and a dense cataract. The intraocular pressure was low. The left cornea showed cornea guttata and incipient cataract. The symptoms ameliorated with NaCl ointment 5 % and later on with contact lens. Combined corneal transplantation and cataract surgery was performed a year later.

The photographs of the right cornea were taken on day 1 (at presentation), day 2 and day 3 and 3 weeks, 5 months and a year later.

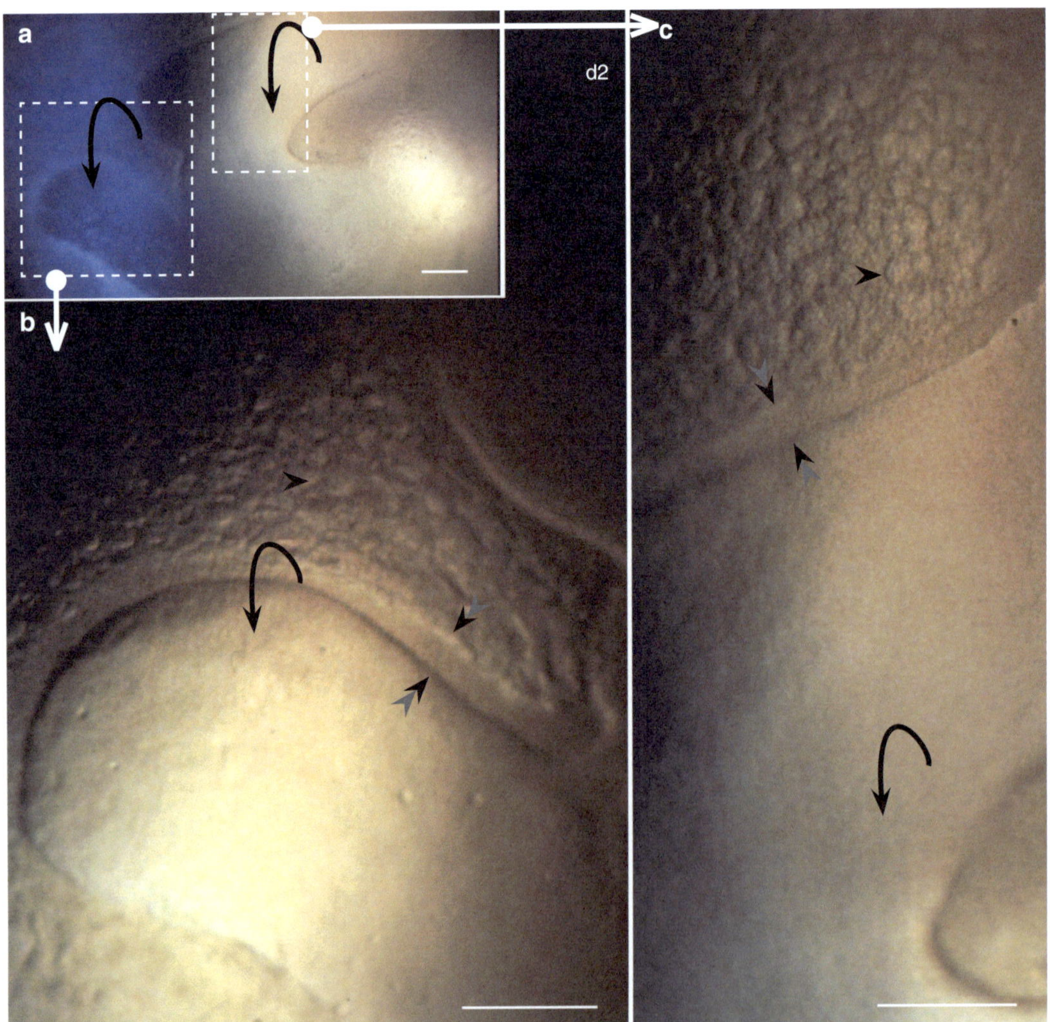

Fig. 5.1 (a) Survey of a part of a very large bulla (*arrows*) (captured on day 2). The areas in *frame* are shown in (b, c): the bulla (*arrows*) is surrounded by a rim (between *double arrowheads*) separating it from epithelium which shows many blisters (*black arrowheads*) and a fine granular pattern of intracellular edema

Case 1. Epithelial Activity in Chronic Edema (cont.)

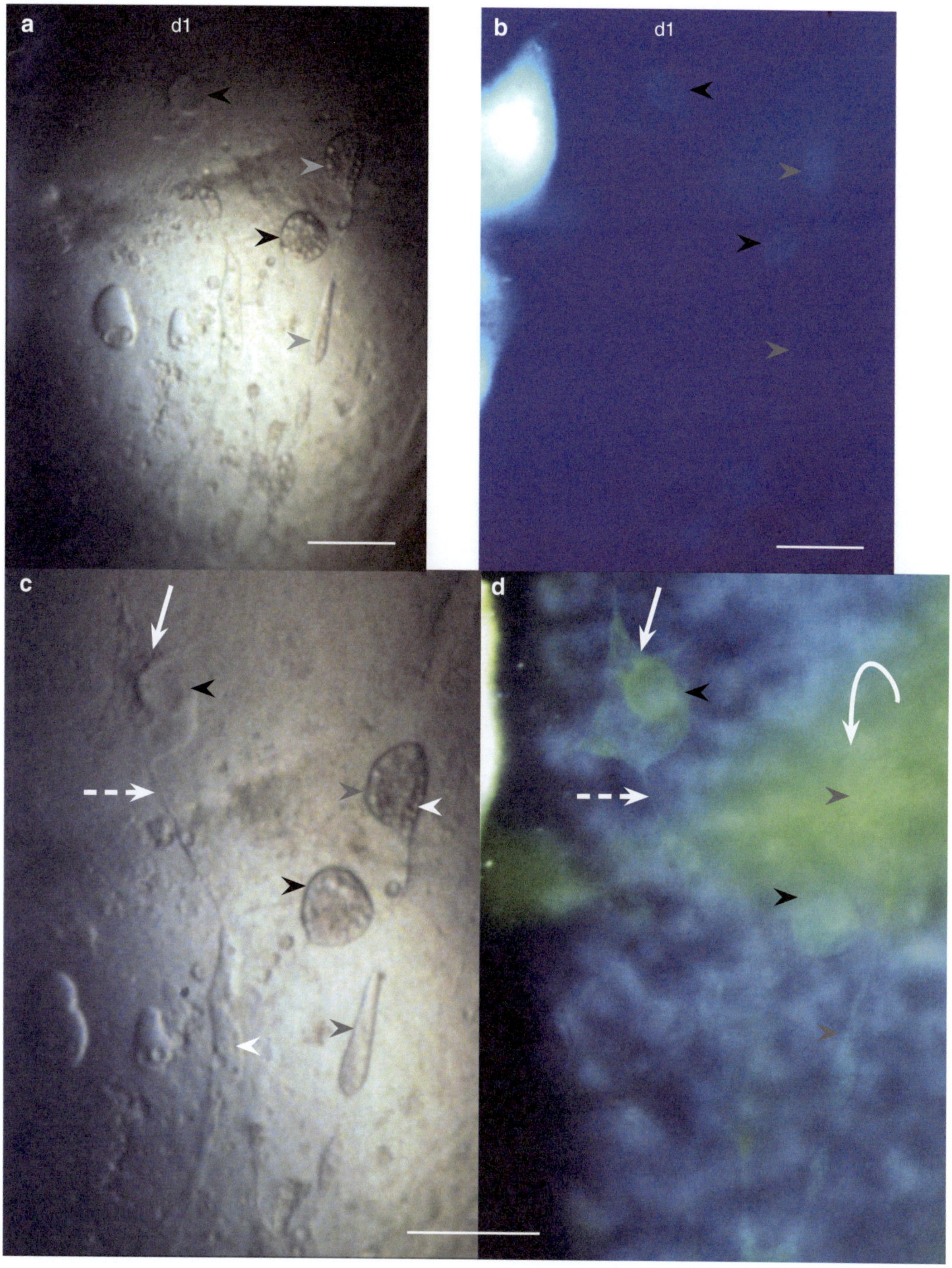

Fig. 5.2 On day 1, in the area of the bulla, are visible (a, c) rounded (*black arrowheads*) and fusiform or pear-shaped (*grey arrowheads*) cystic spaces containing rounded bodies (c, *white arrowheads*); some of the spaces are arranged in rows and some are connected by a thin epithelial line (c, *dashed arrow*). (b, d) Fluorescein visualises some of the cystic spaces (*arrowheads*) and lines (d, *dashed arrow*) and seems to be pooling in a relative surface depression (d, *bowed arrow*). The *plain arrows* in (c, d) indicate a curved line implying an edge of epithelial sheet (the black and grey arrowheads are placed in corresponding locations in all 4 photographs; *b* blue filter)

Case 1. Epithelial Activity in Chronic Edema (cont.)

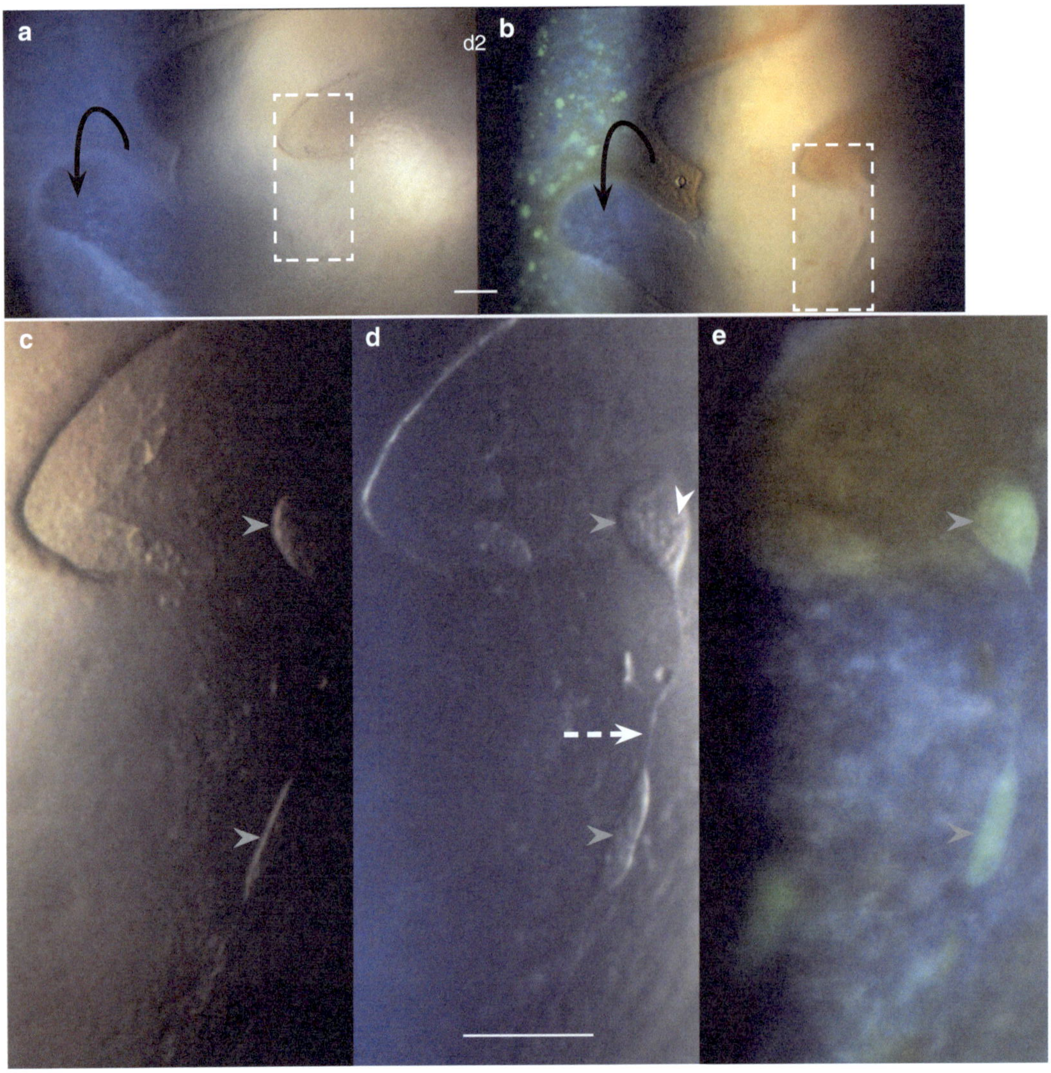

Fig. 5.3 (**a**, **b**) Survey of a large bulla (*arrows*), (**a**) before and (**b**) after the application of fluorescein. In the area in *frames*, shown in (**c**–**e**), a pear-shaped and a fusiform cystic space (*grey arrowheads*) connected by an epithelial line (**d**, *arrow*) is visible. The cystic spaces contain rounded bodies (**d**, *white arrowhead*) and (**e**) stain green with fluorescein (the grey arrowheads are placed in corresponding locations)

Case 1. Epithelial Activity in Chronic Edema (cont.)

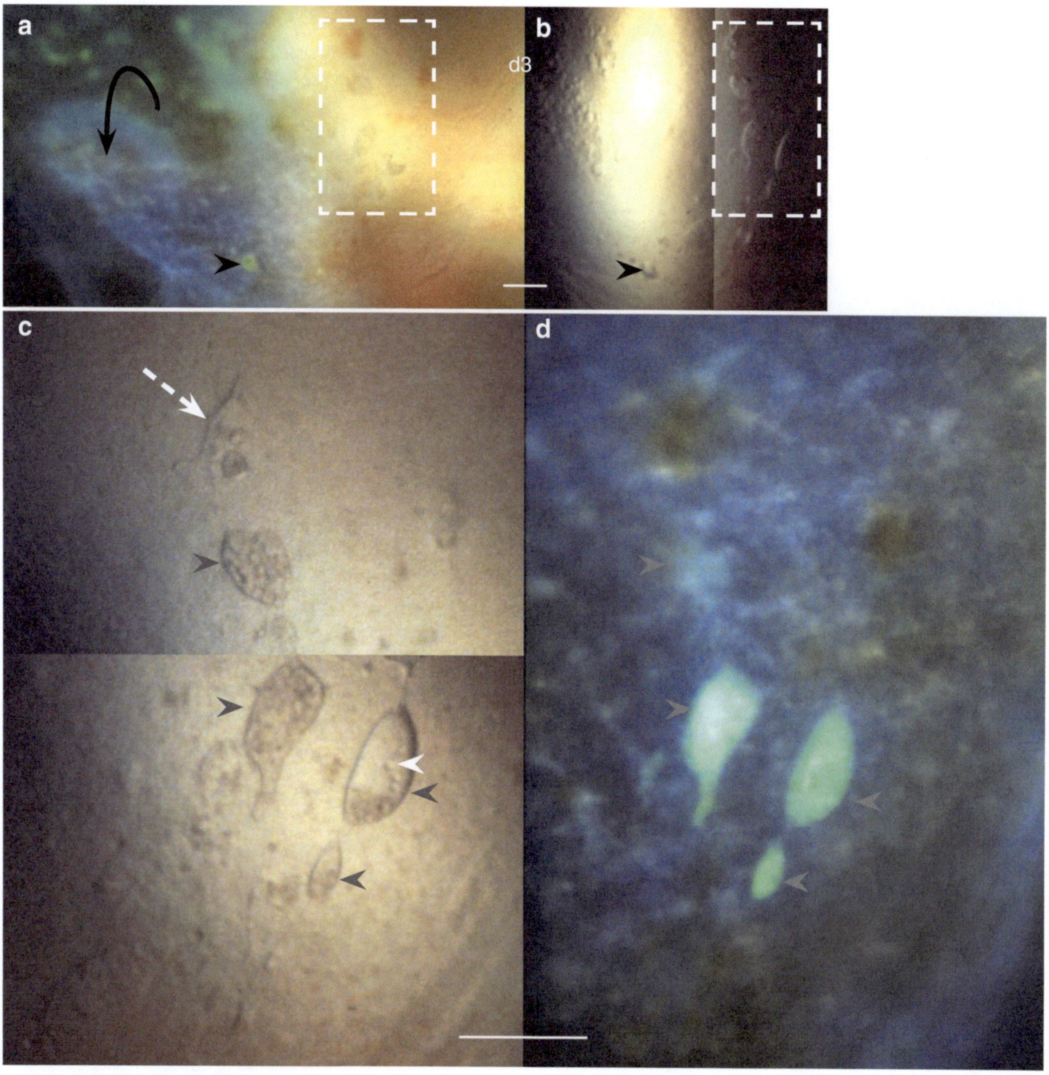

Fig. 5.4 The same large bulla (*arrow*) as in Fig. 5.3. (**a**, **b**) Survey. For orientation, the *arrowheads* indicate the same cystic space. The area in *frames* shows (**c**) variously shaped cystic spaces (*grey arrowheads*) containing rounded bodies (*white arrowhead*) and an epithelial line (*arrow*). (**d**) Some of the cystic spaces stain green with fluorescein (the grey arrowheads are placed in corresponding locations; (**c**) is a composed photograph)

Case 1. Epithelial Activity in Chronic Edema (cont.)

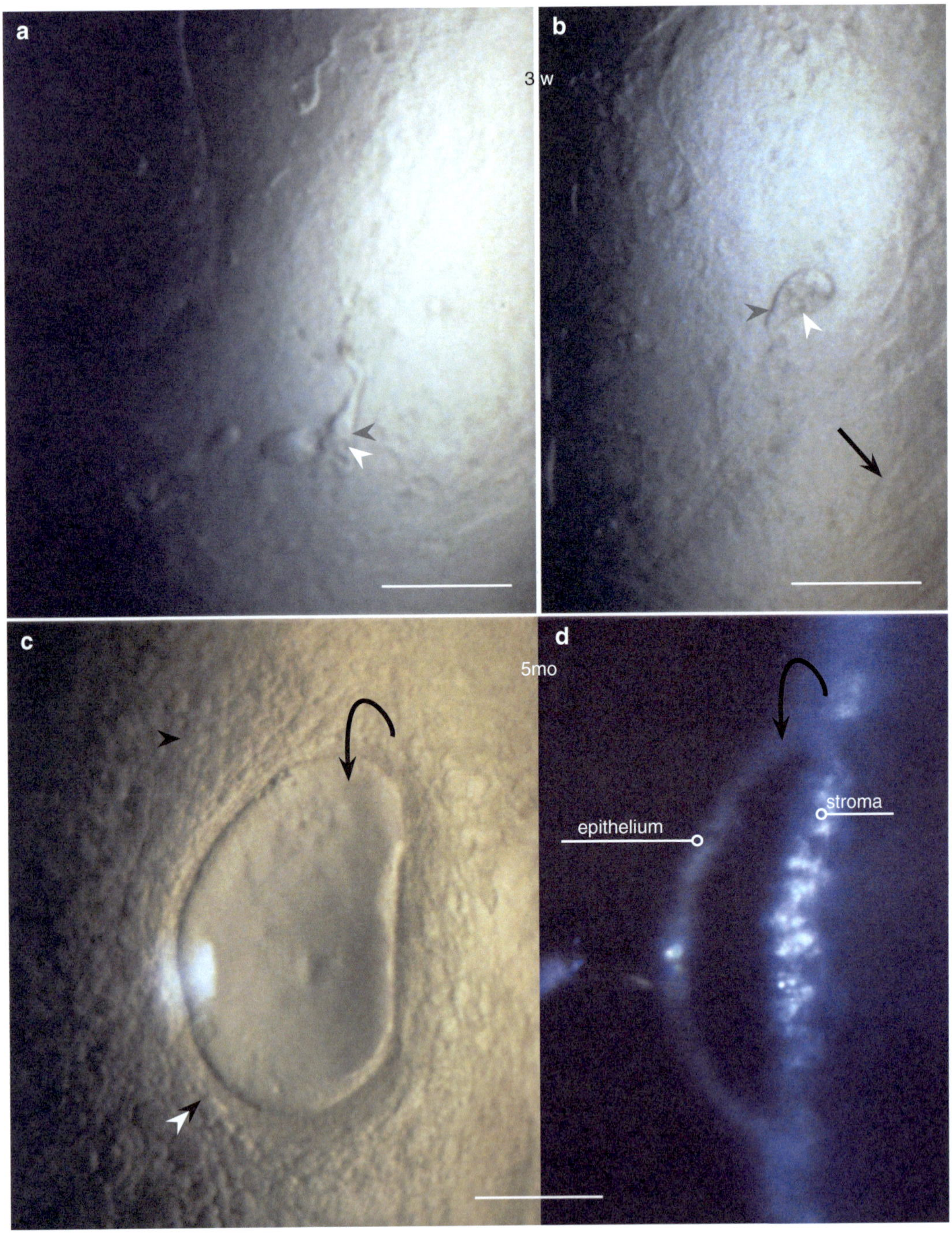

Fig. 5.5 (**a**, **b**) Pear-shaped cystic spaces (*grey arrowheads*), containing rounded bodies (*white arrowheads*), and epithelial wrinkles (**b**, *arrow*). (**c**) A large bulla (*arrow*) situated in an area showing many small blisters (*arrowhead*) which seem missing within the rim adjacent to the bulla (*double arrowhead*). (**d**) Epithelial detachment captured in the bulla (*arrow*) shown in (**c**); its contents are clear

Case 1. Epithelial Activity in Chronic Edema (cont.)

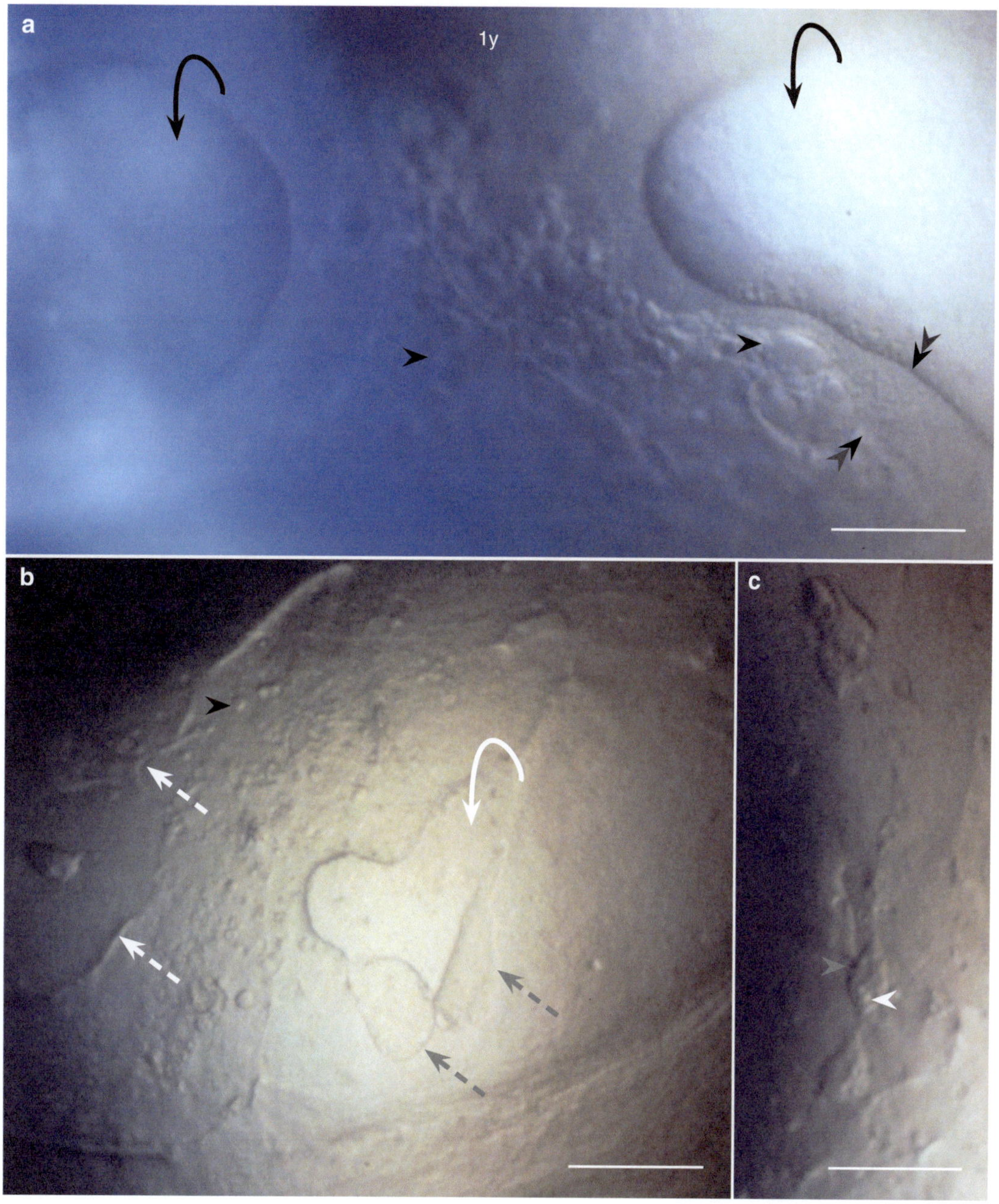

Fig. 5.6 (**a**) Two large bullae (*arrows*) situated in an area showing many blisters (*arrowheads*). Between the *double arrowheads* a relatively smooth rim surrounding the bulla is visible (cf. Figs. 4.13 and 5.1). (**b**) A heart-shaped area (*bowed arrow*) suggestive of epithelial disruption with displaced epithelial flaps (*grey dashed arrows*) at its lower edge. The surroundings show rounded cysts (*arrowhead*) and epithelial lines (*white dashed arrows*) (cf. Fig. 4.10). (**c**) Fusiform cystic spaces (*grey arrowhead*) containing rounded bodies (*white arrowhead*)

Case 2. The Dynamics of Granular Material ('Opaque Cysts') (1)

Case Report

An 80-year-old woman with irritation and redness in the right eye for about 2 weeks was initially treated by general practitioner with antibiotic eye drops. She presented a week later because of augmenting pain and worsening of vision. The lids were swollen, the eye injected, the cornea edematous and the intraocular pressure high (about 60 mmHg), but there was no mydriasis and the depth of the anterior chamber seemed normal. With acetazolamide p.o., within a couple of hours, the pressure dropped and the corneal edema diminished, but the cornea remained hazy. Within short, however, flare in the anterior chamber became discernible and topical corticosteroid was added.

The photographs of the right cornea, taken 3 weeks after presentation (here termed day 1) and 1 and 5 weeks later, captured the development in the paracentral lower cornea of a peculiar figure containing patches of granular material.

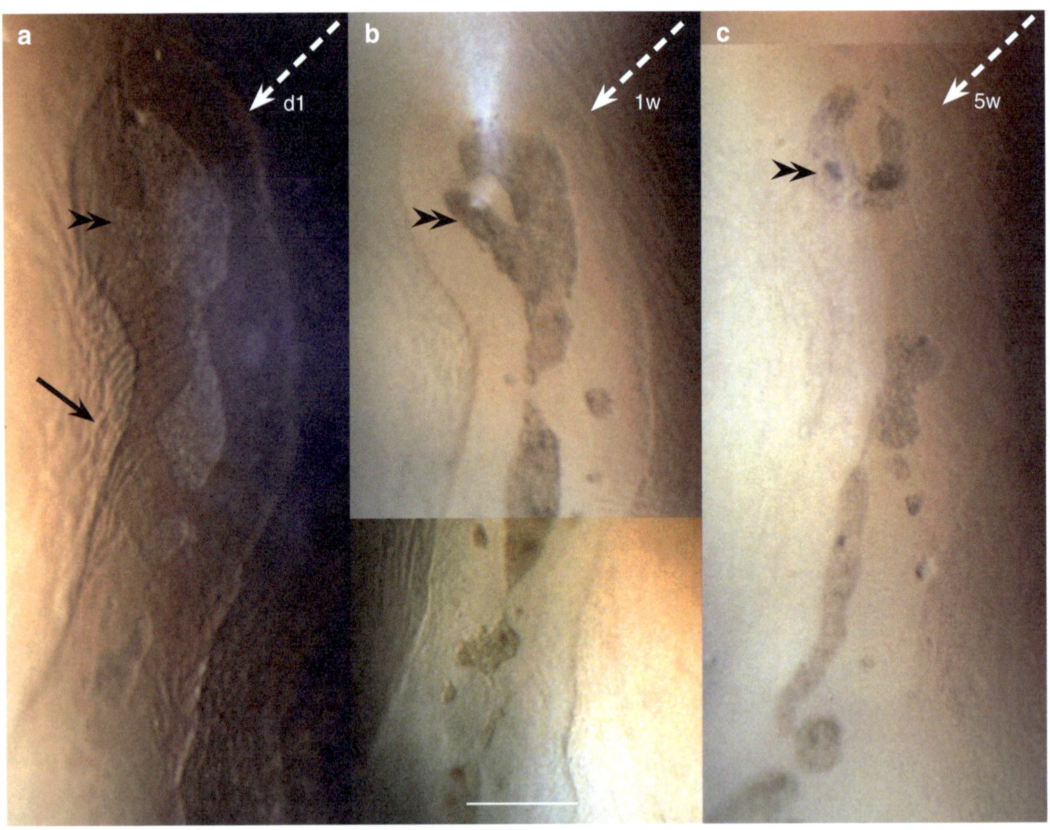

Fig. 5.7 (**a**–**c**) A large, long, well-delineated figure (*white arrows*) containing a row of irregularly shaped patches of granular material (*arrowheads*). The patches change shapes between the occasions. In (**a**) epithelial wrinkles are additionally visible (*black arrow*); cf. Fig. 5.8. ((**b**) is a composed photograph)

Case 2. The Dynamics of Granular Material ('Opaque Cysts') (1) (cont.)

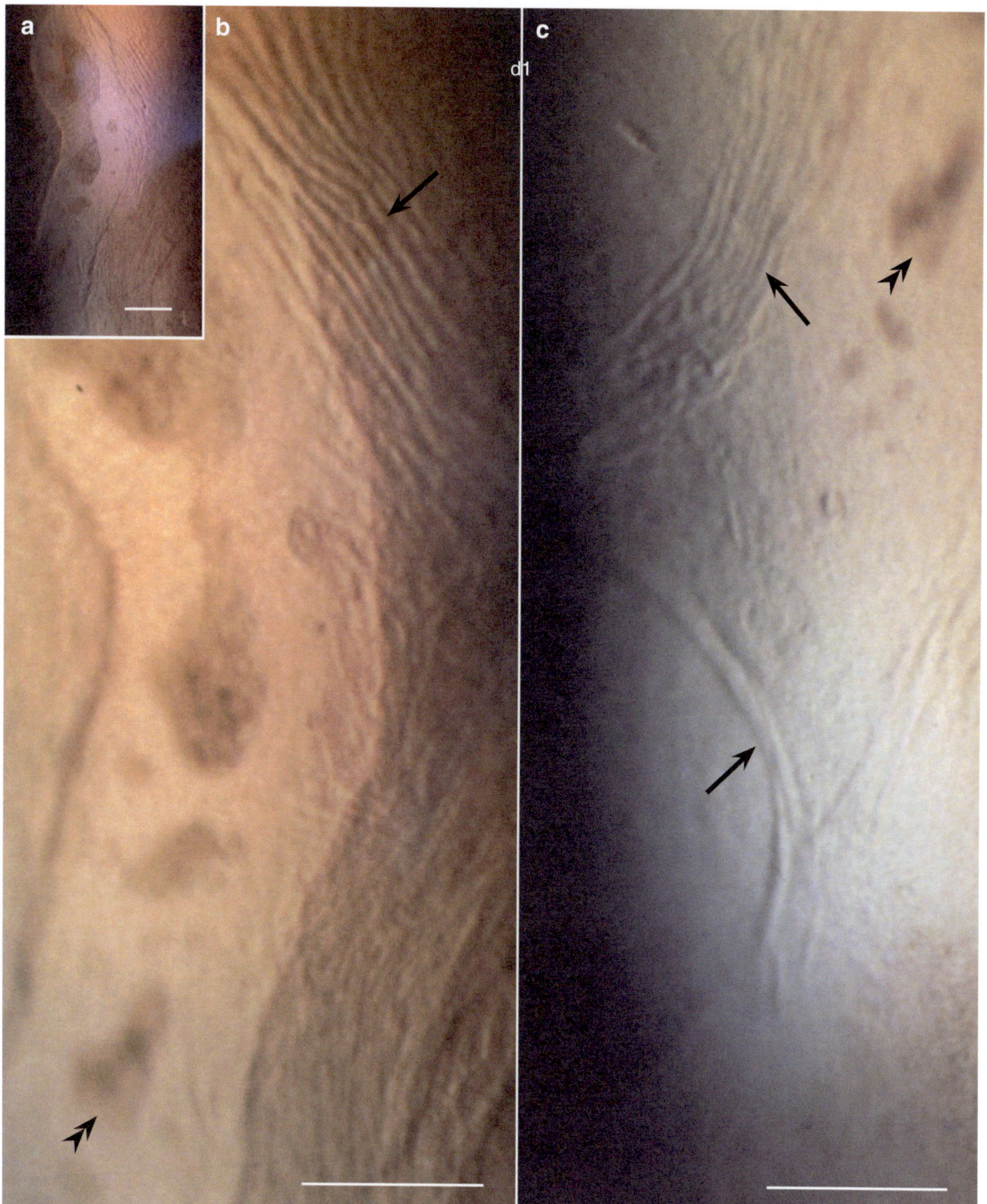

Fig. 5.8 (**a**) Survey of the figure shown in Fig. 5.7a. (**b**) In focus are epithelial wrinkles (*black arrows*) located at its right edge and (**c**) at its lower end. For orientation, the *arrowheads* in (**b**, **c**) indicate the same patch of granular material (here out of focus)

Case 2. The Dynamics of Granular Material ('Opaque Cysts') (1) (cont.)

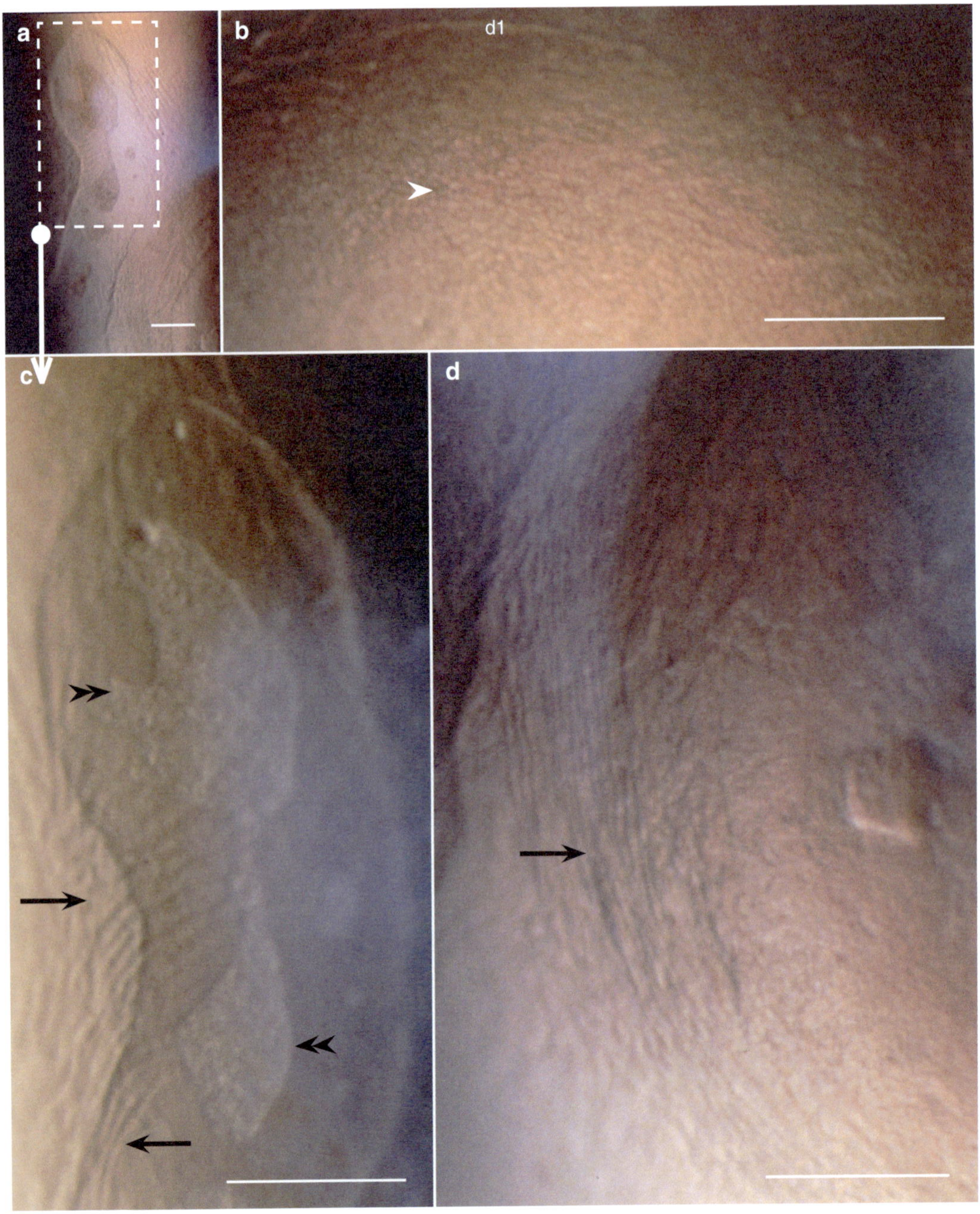

Fig. 5.9 (a) Survey. The area in *frame* is shown in (c). In (b) fine granular pattern of intracellular edema (*arrowhead*) captured outside the figure is visible, in (c) granular contents of the patches (*arrowheads*) within the figure and epithelial wrinkles (*arrows*) extending over the figure and in (d) epithelial wrinkles (*arrow*) located at the figure's right lower edge

Case 2. The Dynamics of Granular Material ('Opaque Cysts') (1) (cont.)

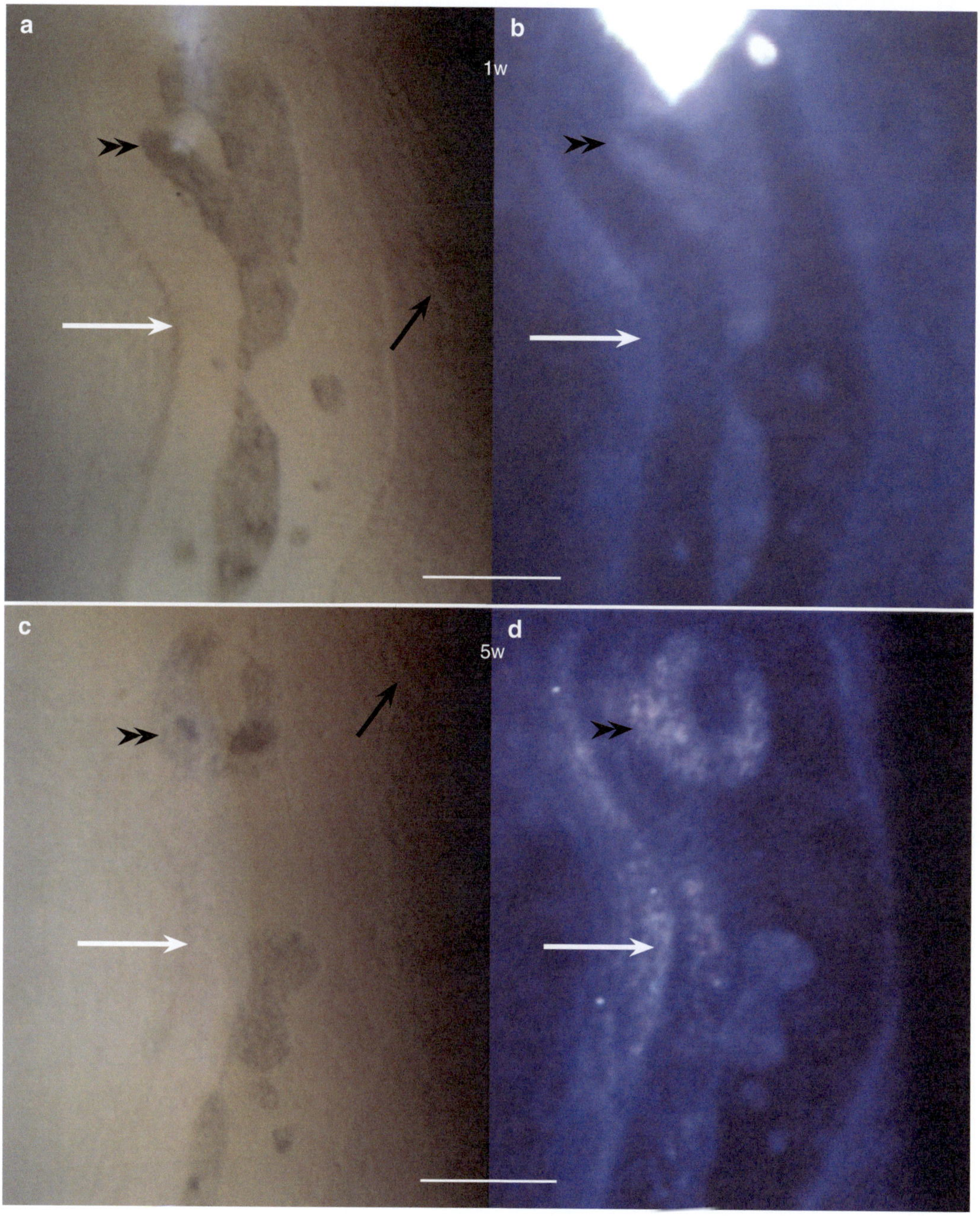

Fig. 5.10 (a, c) The patches of granular material (*arrowheads*) are (b, d) light-reflecting and so are the edges of the figure (*white arrows*). The *black arrows* indicate epithelial wrinkles (the markers are placed in corresponding locations in each pair of photographs)

Comment

A suspicion of herpes simplex as the cause was supported by the patient's history of recurrent herpes labialis and a protracted inflammation of unknown origin in one eye, both several years previously. The cornea itself did not show any changes diagnostic of herpes simplex.

Case 3. The Dynamics of Granular Material ('Opaque Cysts') (2)

Case Report

An 81-year-old woman with bilateral cornea guttata, cataract and, in the left eye, a history of acute glaucoma and cataract surgery with superior incision and IOL.

The photographs of the left upper cornea were taken 2 months after the surgery (here termed day 1) and 2 weeks, 3 months and 1 year later.

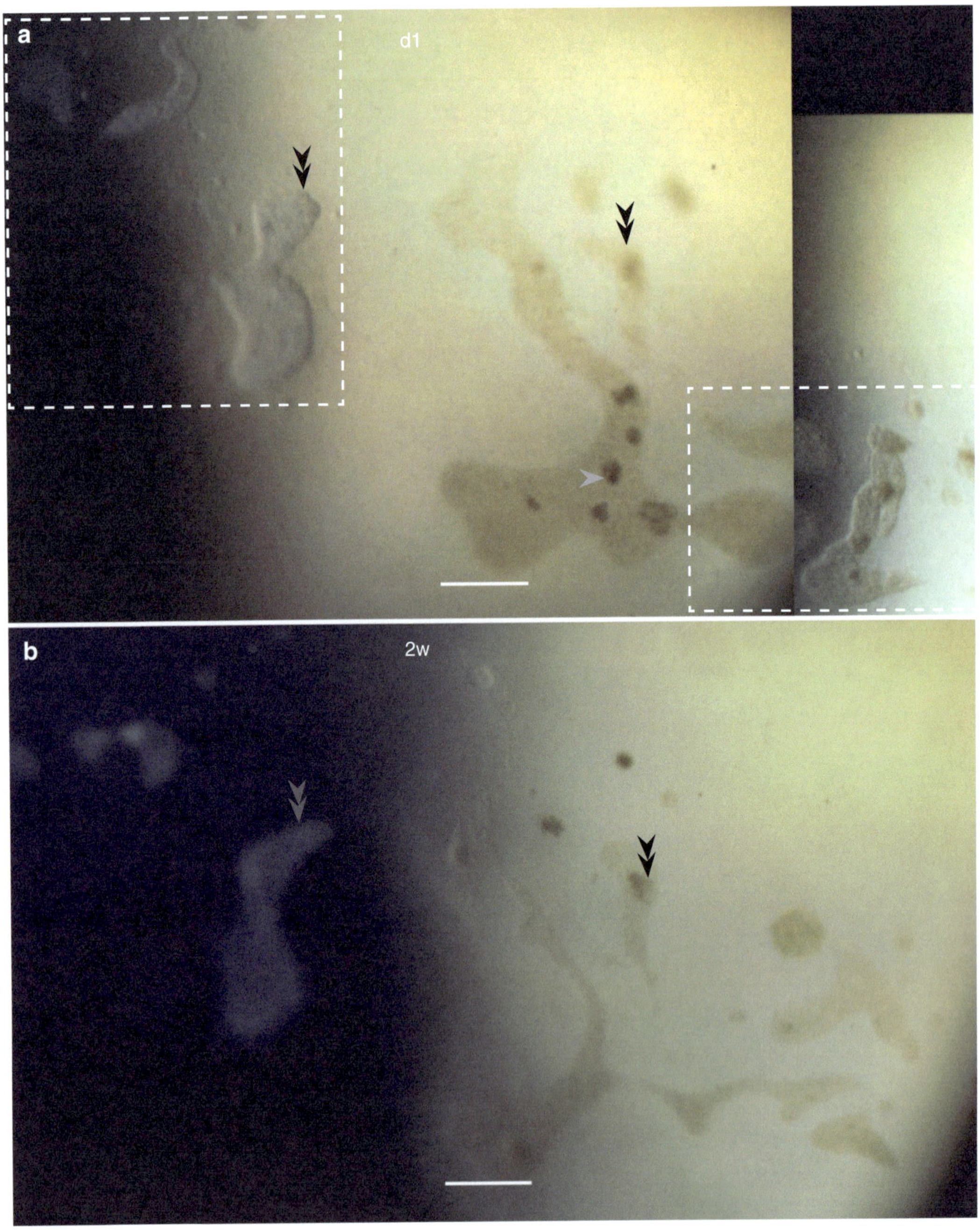

Case 3. The Dynamics of Granular Material ('Opaque Cysts') (2) (cont.)

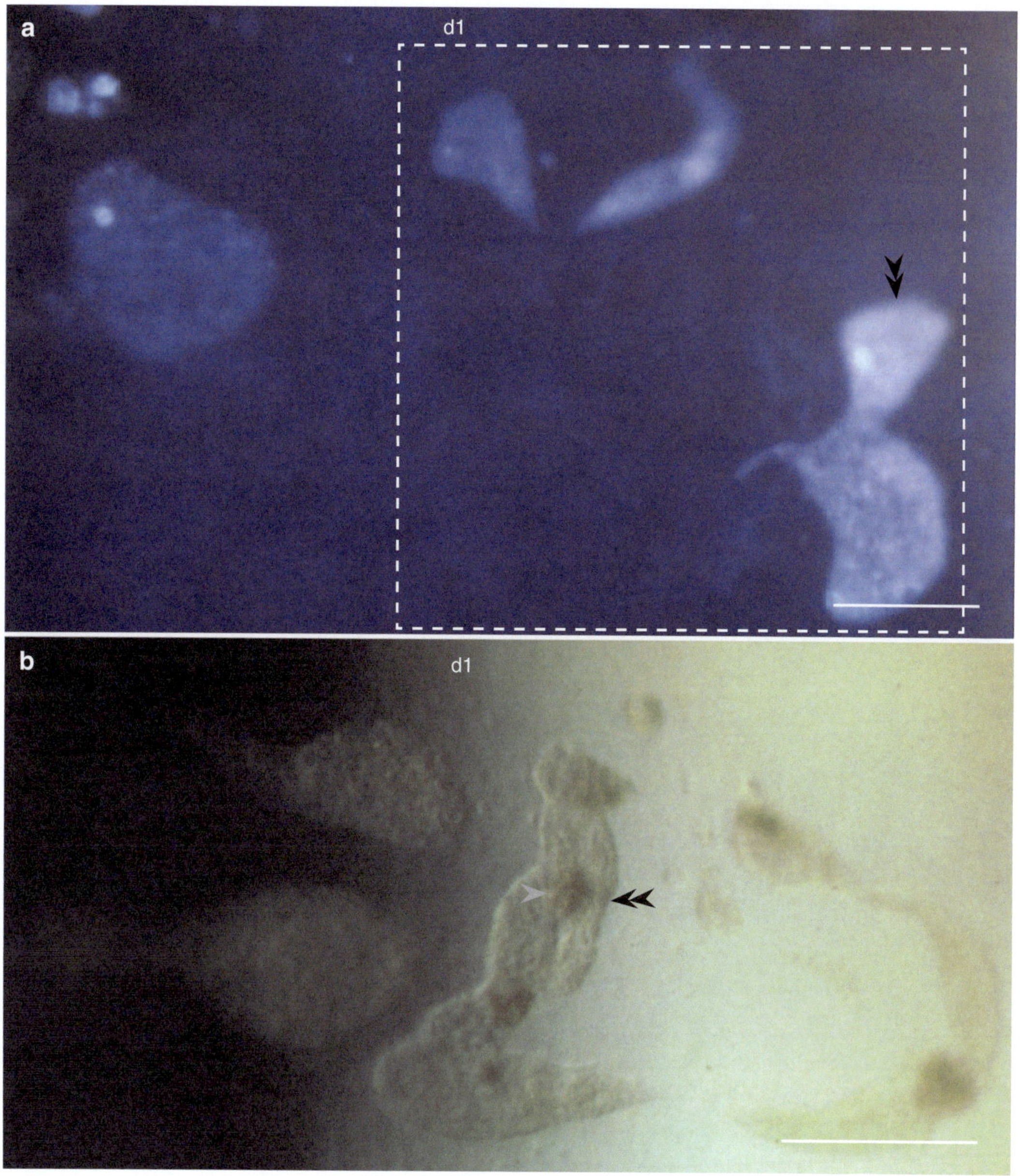

Fig. 5.12 (**a**) The figures and patches (*double arrowhead*) are strongly light-reflecting, cf. the area in left frame in Fig. 5.11a (the arrowheads are placed in corresponding locations). (**b**) shows the granular appearance of the figures and patches (*double arrowhead*) and denser (darker) spots within them (*arrowhead*). For localisation cf. right frame in Fig. 5.11a

Fig. 5.11 Survey. (**a**) Granular material forming variously shaped figures and patches (*double arrowheads*) of which some contain denser spots (*arrowhead*). The areas in *frames* in (**a**) are shown in Fig. 5.12. (**b**) Within 2 weeks, the figures have changed shapes: some parts have disappeared and new ones developed. Reference points are lacking, but some locations are recognisable. (The double arrowheads are placed in corresponding locations; (**a**) is a composed photograph.) Compare also Figs. 1.6 and 1.7

Case 3. The Dynamics of Granular Material ('Opaque Cysts') (2) (cont.)

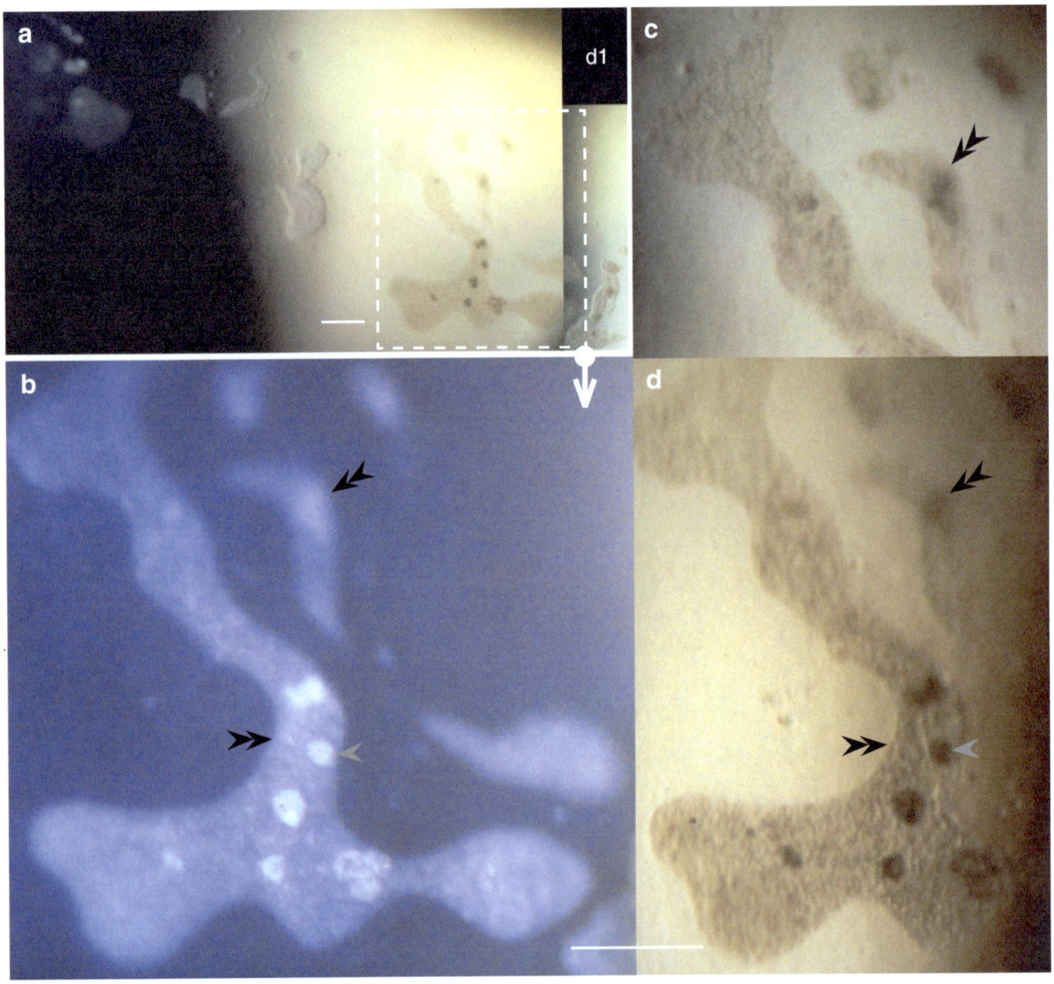

Fig. 5.13 (a) Survey. (b–d) show the area in *frame* at higher magnification. (b) The light-reflecting figures (*double arrowheads*) contain brighter spots (*arrowhead*). In (c) the corresponding upper part and in (d) the lower part of the area are in focus. Both appear granular; the denser (darker) spots (*arrowheads*) correspond to the bright ones in (b) (the markers are placed in corresponding locations) (see Fig. 5.14 showing the same area 2 weeks later)

Case 3. The Dynamics of Granular Material ('Opaque Cysts') (2) (cont.)

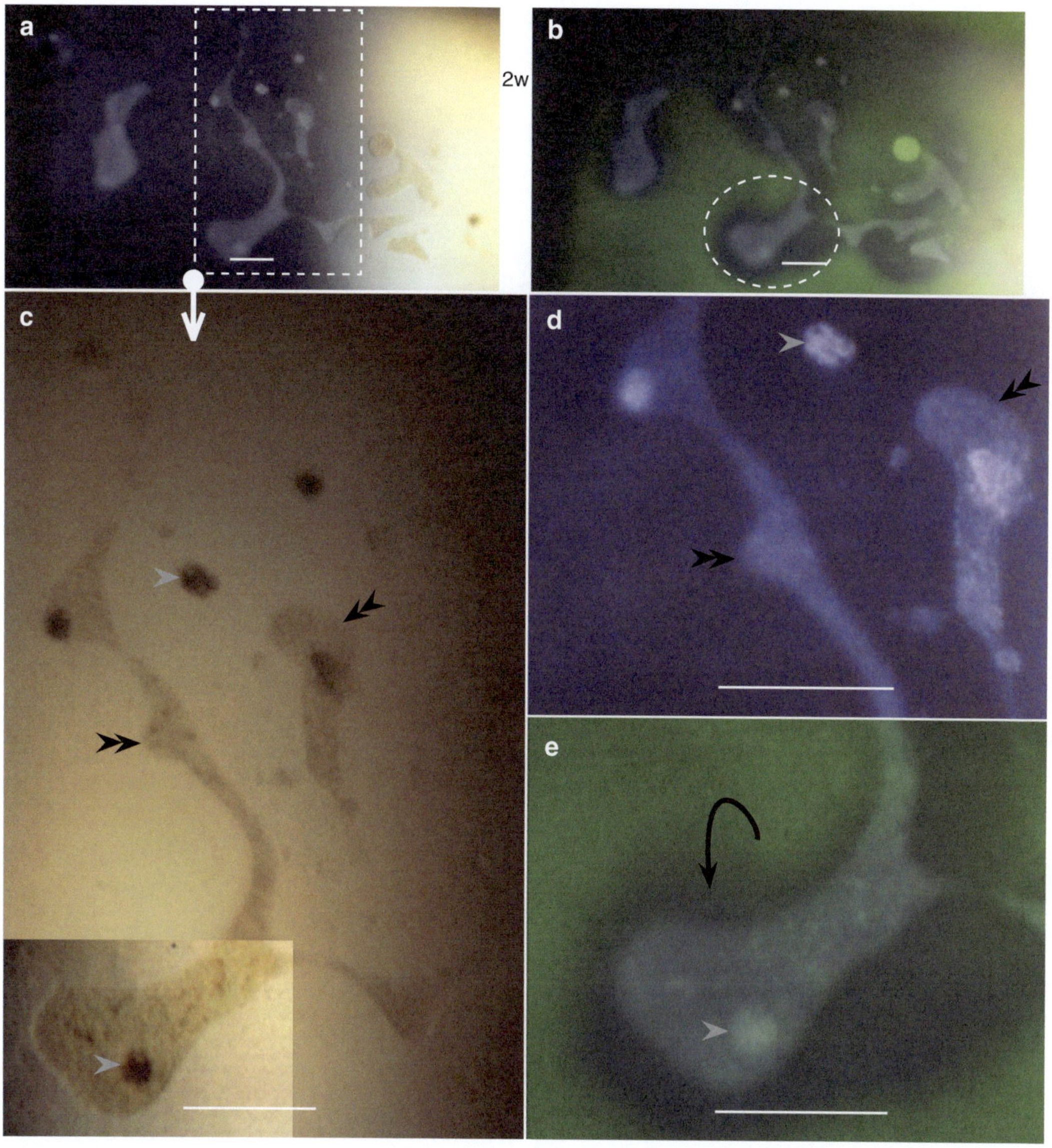

Fig. 5.14 Survey. (**a**) Before and (**b**) after fluorescein staining. In the green stained tear film are visible surface elevations (dark) surrounding the figures; the area in *oval frame* is shown in (**e**). (**c**) shows denser (darker) spots (*arrowheads*) both within and outside the figures (*double arrowheads*); the figure to the right (*double arrowhead*) is present approximately in the same location as before (cf. Figs. 5.11 and 5.13). In (**d**) is visible the variously strong light-reflecting property of the figures and in (**e**) surface elevation (*arrow*) around the lower part of the large figure the *arrowhead* indicates the same spot as in (**c**) (the arrowheads are placed in corresponding locations)

Case 3. The Dynamics of Granular Material ('Opaque Cysts') (2) (cont.)

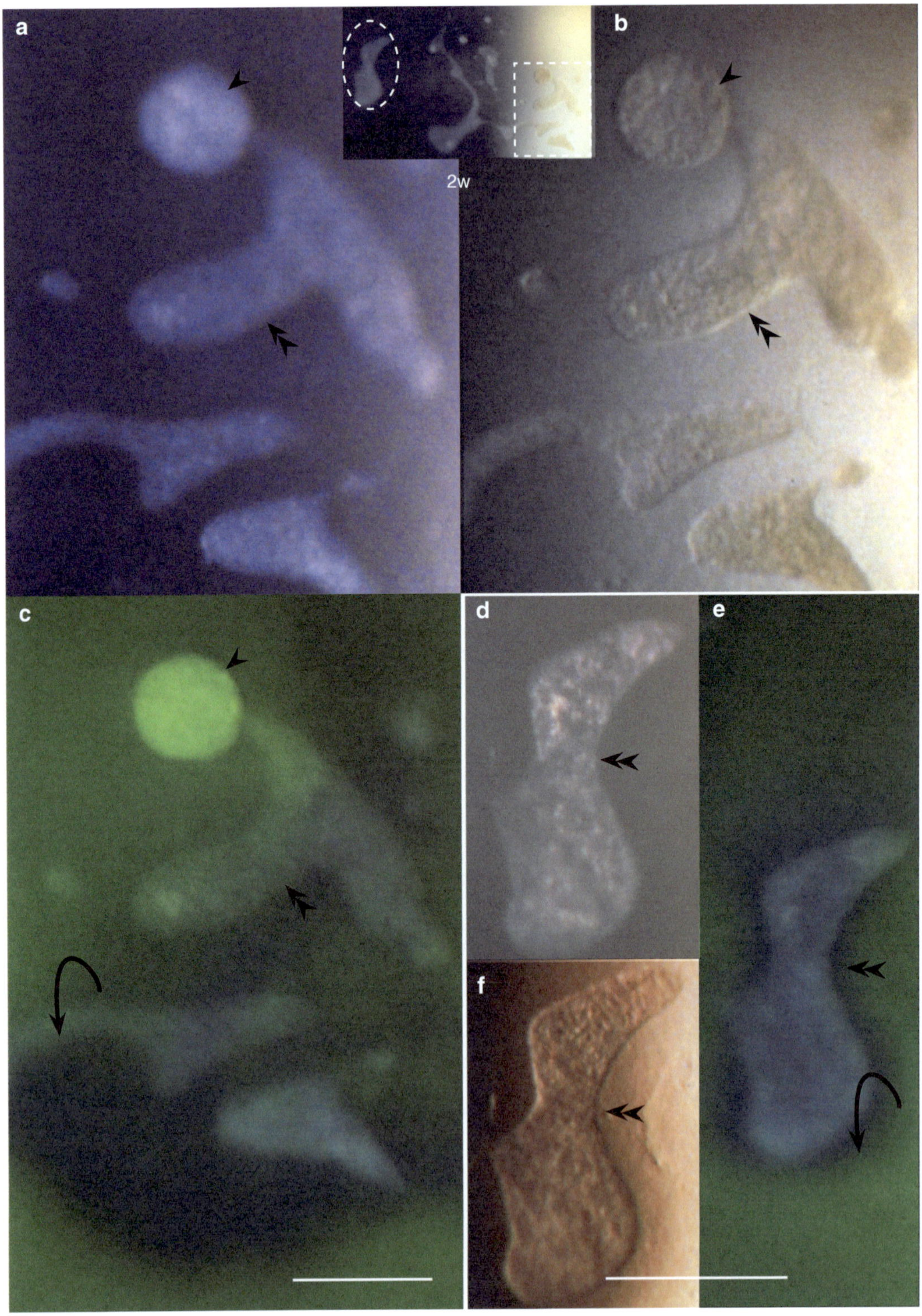

Case 3. The Dynamics of Granular Material ('Opaque Cysts') (2) (cont.)

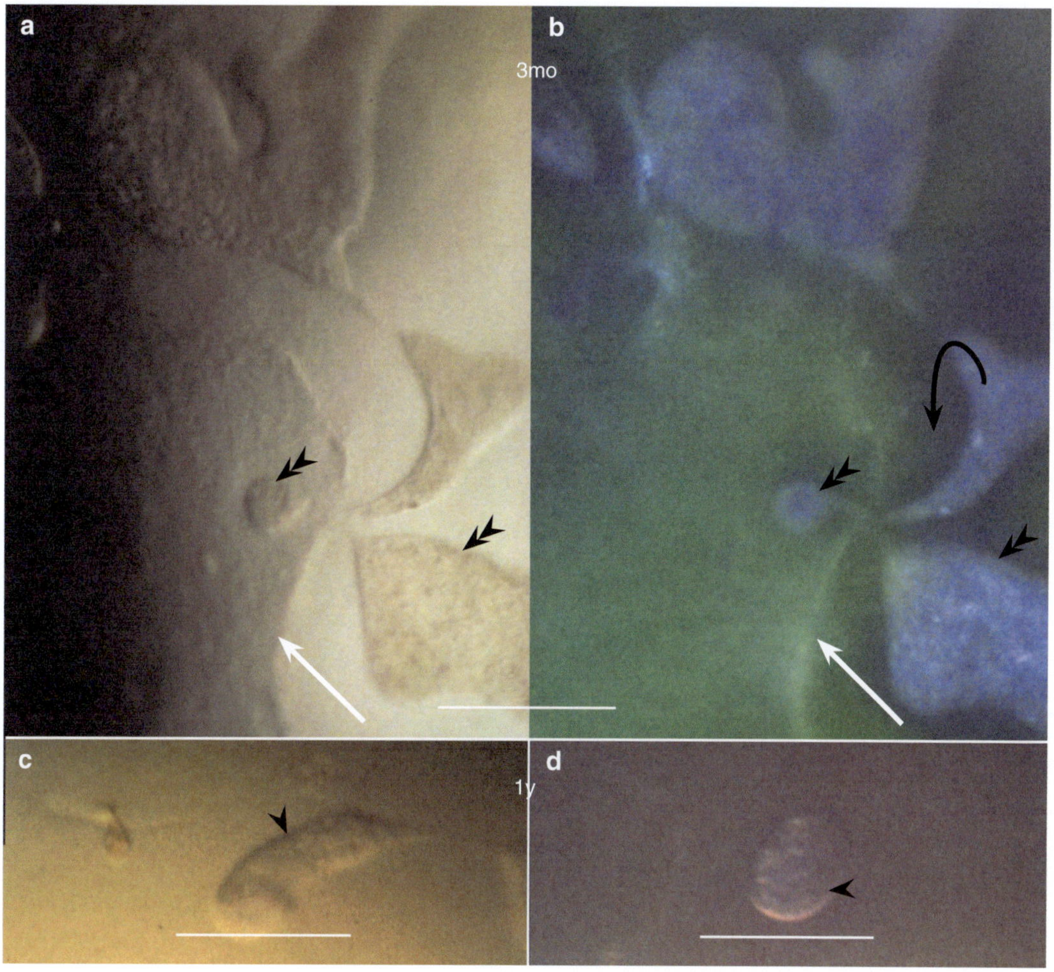

Fig. 5.16 (a) An area showing several figures (*double arrowheads*) in apposition; the granularity of the contents are visible in the upper one. (b) The figures are light-reflecting except for one (*straight arrow*) that has disappeared under the tear film layer stained green with fluorescein. The *bowed arrow* indicates surface elevation (dark). (The arrowheads and the straight arrows are placed in corresponding locations.) (c, d) Only few small figures/cyst-like lesions (*arrowheads*) were present at this occasion

Fig. 5.15 *Inset*: Survey. The area in *rectangular frame* is shown in (a–c) and that in *oval frame* in (d–f). (a–c) The perfectly rounded part (*arrowheads*) of this (a) light-reflecting figure (*double arrowheads*) with (b) granular contents stains (c) brilliantly green with fluorescein; the *arrow* in (c) indicates surface elevation (dark). (d–f) The figure (*double arrowheads*) indicated by oval frame in the inset appears granular; the dark surrounding zone (*arrow*) indicates surface elevation (the arrowheads are placed in corresponding locations in each set of photographs)

Final Remark

The well-established term *recurrent erosion* implies epithelial erosion in its classic meaning, i.e. a surface disruption with loss of substance. In patients with symptoms pointing to the syndrome, however, frank erosions are considerably less often observed than epithelial derangement showing larger and/or smaller areas of disturbance, the irregular shapes of which defy description, often in combination with cysts and rounded bodies spread in the surroundings.

Rounded bodies are almost always present. With the present method no distinction can be made between the origins of the *light-reflecting* ones, i.e. between degenerating epithelial cells with altered contents and invading inflammatory cells. The *not light-reflecting* rounded bodies appearing as dark dots in tear fluid stained green with fluorescein, on the other hand, most probably belong to a different category, i.e. swelling cells with clear contents; this phenomenon occurs also in other conditions, for example, epithelial virus infections. Round *cystic spaces* with more or less clear contents surrounding light-reflecting rounded bodies are very common in recurrent erosions and rare in other conditions (own observations). Light-reflecting rounded bodies and/or cell debris are a substantial part of larger areas of disturbance, here termed *lesions*, and they are the main element behind their light-reflecting property. The lesions, with rare exceptions, are sharply delineated; their contents occasionally appear almost uniform, but the majority of them show a broken pattern as if resulting from an asynchronous disease process; some contain also peculiar, well-delineated, circular or oval areas difficult to interpret; and in some are discernible reparative features.

The depth of these changes cannot be estimated in two-dimensional images, but important information can be obtained with the aid of *fluorescein sodium*. Generally, in focal illumination, surface *elevations* appear dark in the green stained tear film, *enlarging green flecks* indicate fluorescein diffusion into the tissues, and the dye is pooling within surface *depressions* and within *(semi)cystic spaces* that have a surface connection.

In the recurrent erosion syndrome, frank erosions excepted, the *absence of fluorescein diffusion* into the surrounding tissues is striking. Not staining rounded cysts and other structures causing surface elevations point to an intraepithelial pathological process as the primary event, and the staining ones to the result of a successive transportation and/or expansion of abnormal structures toward the surface that ultimately starts to disrupt. The persistence of the green staining in circumscript spaces long after the dye had disappeared from the tear film implies *(semi)cystic spaces*, and the successive penetration of the dye captured in some lesions indirectly visualises the presence of an anterior wall and indicates an approximate location of a small surface connection(s). Such connections have been captured in some of the *opaque cysts* that, apparently entrapped below but close to the edge of a membrane ('map'), formed channels through locus minoris resistentiae to empty their contents on the surface (Chap. 2, Case 13). This, however, represented a particular situation. On the whole, opaque cysts are known to only exceptionally

H.M. Tabery, *Recurrent Erosion Syndrome and Epithelial Edema: In Vivo Morphology in the Human Cornea*, DOI 10.1007/978-3-319-06545-8, © Springer International Publishing Switzerland 2015 stain green with fluorescein. Dynamic observations (Chap. 5, Case 3) show that a slow elimination of light-reflecting granular material accumulated in variously shaped opaque cysts (known from histological examinations to consist of degenerated cells/cell debris) indeed can occur without causing clinically detectable surface disruptions. It seems that during their elimination fluid penetration into the cysts either does not occur or perhaps escapes detection because of the light-reflecting property of the cysts hampering the visibility of the green fluorescence. A slow remodelling of the cysts is also implied by their bizarre shapes.

The corneal epithelium is a highly *dynamic tissue* in which younger cells perpetually substitute the oldest ones shed from the surface. How the presence of abnormal intraepithelial elements influences their turnover, under normal conditions estimated to about 7 days, seems unknown. Isolated small entities, such as a few rounded bodies and rounded cysts, probably disappear by a successive transportation to and shedding from the surface. When concentrated, however, they cause a disturbance eliciting reparative responses. In comparison with well-defined experimental injuries with cell-denuded wound bottom, in which repair starts by epithelial sliding to close the wound, the *repair process* in the recurrent erosion syndrome (frank erosions excepted) differs in several important respects: there is no clearly defined starting point, neither in time nor in location; the surface to be repaired is not cell denuded; and the repair is complicated by a disease process that might or might not have ceased.

Hence, in this complex setting, which are the *features relatable to reparative activity*? Dynamic observations reveal that curved lines at the edges of larger lesions are not only suggestive of but indeed represent *edges or tips of epithelial sheets* more or less successfully attempting to slide over a not cell-denuded surface. They also show variations in *epithelial lines* formed by opposite edges just before merging. Such lines may be short or long; the latter may be in places disrupted by an already restored surface, expanded by fusiform spaces of which some contain not light-reflecting rounded bodies and show development of pear-shaped clear cysts in the path of sheet merging. In sharply delineated lesions, their irregular, wavy edges imply that in places the advancement of epithelial sheets has been slowed down or arrested by the presence of abnormal material, either residual one or one reflecting a contemporaneously ongoing damage. Some smaller lesions may be residua of larger ones already subjected to the repair process; some larger ones show features implying more than one wave of sliding epithelium; and in some lesions rounded, smooth, and *elevated islands* suggest that also an ability of focal epithelial regeneration plays a role in the repair (cf. Chap. 2, Case 9).

Occasionally, patients present with a well-delineated *edematous epithelial area* developing either spontaneously or after a trivial injury. The mechanisms behind this phenomenon are difficult to deduce. The follow-up capturing one such lesion (Chap. 2, Case 10) shows that the affected epithelium does not necessarily slough off as a whole. During consecutive 3 days, curved epithelial edges advanced toward each other apparently by sliding on top of some preserved epithelial layer(s), and the initially present large numbers of not light-reflecting rounded bodies suggestive of swollen cells successively disappeared. Five days after injury, the involved area was considerably smaller but still recognisable mainly because of a not yet completed merging of opposite edges and of level differences implying areas of epithelial hyperplasy.

Contrary to the rapid changes occurring in the epithelium, phenomena relatable to abnormalities of basement membrane and/or deposits of abnormal material (maps) change very slowly (Chap. 2, Case 13).

Recurrent erosions are very common but only rarely complicated by *stromal infiltrates*. The follow-up performed in two patients showed the same features: after 3 days of treatment with topical steroid, the infiltrates appeared as superficial, rounded or lobulated whitish opacities containing dense rounded spots, in one rounded infiltrate perfectly located in the centre, and it was here that the clearing of the infiltrates started. Possibly, these spots visualised points of entry of an agent attracting inflammatory cells. Whatever the nature of the agent, there seems to be a causal relationship between recurrent erosions and apparently sterile stromal infiltrates.

Epithelial edema, in addition to its typical features, occasionally shows phenomena similar to or indistinguishable from those occurring in the recurrent erosion syndrome such as *epithelial lines and cystic expansions* implying reparative activity (Chap. 5, Case 1). *Opaque cysts*, developing in edematous corneae (Chap. 5, Case 2) or as a sequela of (presumed) corneal edema close to corneal incision in cataract surgery (Fig. 1.7 and Chap. 5, Case 3), resemble those occurring in EBMD. Because they are rare, speculations on possible links between the two diseases seem fruitless, but they might be confusing in clinical practice.

Recurrent erosion syndrome is occasionally *confused with* other diseases, sometimes with dry eye but more often with herpes simplex virus infections, probably because both recur and show epithelial 'blisters'. Yet, their morphology is entirely different, and a meticulous observation usually resolves the problem. Filaments seem rather rare.

On the whole, the corneal epithelium in recurrent erosion syndrome shows an abnormal activity in conjunction with preserved natural reparative propensity, which, in clinical practice, seems aided by simple means. So far, however, the causes of and mechanisms behind the abnormal epithelial activity are not clear and a treatment in the long term preventing recurrences remains elusive.

Bibliography

Tabery HM (1992) Dual appearance of fluorescein staining in vivo of diseased human corneal epithelium. A non-contact photomicrographic study. Br J Ophthalmol 76:43–44

Tabery HM (1995) Morphology of herpes simplex dendritic keratitis. A non-contact photomicrographic study in the human cornea. Herpes 2:55–57

Tabery HM (1997) Micropunctate staining of the human corneal surface: microerosions or cystic spaces? A non-contact photomicrographic in vivo study. Acta Ophthalmol Scand 75:134–136

Tabery HM (1998) Corneal stromal infiltrates in patients with recurrent erosions. Acta Ophthalmol Scand 76:589–592

Tabery HM (2003a) Corneal surface changes in keratoconjunctivitis sicca. Part I: the surface proper. A non-contact photomicrographic in vivo study in the human cornea. Eye 17:482–487

Tabery HM (2003b) Corneal surface changes in keratoconjunctivitis sicca. Part II: the mucus component. A non-contact photomicrographic in vivo study in the human cornea. Eye 17:488–491

Tabery HM (2003c) Filamentary keratopathy: a non-contact photomicrographic in vivo study in the human cornea. Eur J Ophthalmol 13(7):599–605

Tabery HM (2004) Corneal surface changes in Thygeson's superficial punctate keratitis: a clinical and non-contact photomicrographic in vivo study in the human cornea. Eur J Ophthalmol 14:85–93

Tabery HM (2010) Herpes simplex virus epithelial keratitis. In vivo morphology in the human cornea. Springer, Berlin/Heidelberg

Tabery HM (2011) Varicella-zoster virus epithelial keratitis in herpes zoster ophthalmicus. In vivo morphology in the human cornea. Springer, Berlin/Heidelberg

Tabery HM (2012a) Adenovirus epithelial keratitis and Thygeson's superficial punctate keratitis. In vivo morphology in the human cornea. Springer, Berlin/Heidelberg

Tabery HM (2012b) Keratoconjunctivitis sicca and filamentary keratopathy. In vivo morphology in the human cornea and conjunctiva. Springer, Heidelberg/New York/Dordrecht/London

Tabery HM, Holm O (1987) Photography in vivo of epithelial lesions in the human cornea. Acta Ophthalmol (Kbh) 65:513–514

DOI 10.1007/978-3-319-06545-8,

Index

H.M. Tabery, *Recurrent Erosion Syndrome and Epithelial Edema: In Vivo Morphology in the Human Cornea*,
DOI 10.1007/978-3-319-06545-8,

MIX
Papier aus verantwortungsvollen Quellen
Paper from responsible sources
FSC® C105338

If you have any concerns about our products,
you can contact us on
ProductSafety@springernature.com

In case Publisher is established outside the EU,
the EU authorized representative is:
Springer Nature Customer Service Center GmbH
Europaplatz 3, 69115 Heidelberg, Germany

Printed by Libri Plureos GmbH
in Hamburg, Germany